LAND OF
LEAD

*To Ann for her patience and support,
and Eleri and Seren for the future*

LAND OF LEAD

THE STORY OF FOUR GENERATIONS
OF A NORTH CEREDIGION FAMILY

BRIAN DAVIES

First impression: 2021

© Copyright Brian Davies and Y Lolfa Cyf., 2021

The contents of this book are subject to copyright, and may
not be reproduced by any means, mechanical or electronic,
without the prior, written consent of the publishers.

Cover design: Y Lolfa
Cover image: Brian Davies

ISBN: 978 1 78461 966 4

Published and printed in Wales
on paper from well-maintained forests by
Y Lolfa Cyf., Talybont, Ceredigion SY24 5HE
website www.ylolfa.com
e-mail ylolfa@ylolfa.com
tel 01970 832 304
fax 832 782

Painting of a schooner in full sail by Captain William James

Contents

Preface

BOTH SIDES OF my family, maternal and paternal, are deeply rooted in the Aberystwyth area of north Cardiganshire, nowadays known as Ceredigion. My maternal great-great-grandfather was a sailing ship sea captain operating out of Aberystwyth, as was his son, my great-grandfather. My paternal grandfather and great-grandfather were lead miners employed in the mines in the hills surrounding the town.

I am fortunate that a wealth of both written and pictorial records survived from them. These provide a historical insight into life in north Cardiganshire from the 19th century to the Sixties of the 20th century. The interconnected links of my family – from the decline of maritime activity and the lead mining industry to the establishment of rail links, especially the narrow gauge Vale of Rheidol Branch – provide a detailed historical commemoration of events which impacted greatly on the area. These changing times were further influenced by the effect of two world wars.

This book recalls changing, yet turbulent times in Aberystwyth and the surrounding rural area.

Brian Davies
April 2021

Introduction

AT THE CENTRE of the long coastal curve of Cardigan Bay lies Aberystwyth, a historical municipality whose history as an established township can be traced back to 1277 when it was granted free borough status by King Edward I. People had lived in the area for thousands of years prior to this and the impressive conical hill of Pen Dinas, overlooking the present town to the south, gives testimony to the ancient occupation of the region.

Surrounding the upper section of the hill are significant earth ramparts which were part of an ancient hill fort dating back to the Bronze and Iron Ages, where sophisticated and well-organised communities lived between 1200 BC and 600 BC.

The settlement of Aberystwyth was built as a walled town by Edmund of Lancaster, known as Edmund Crouchback, the brother of King Edward I who was also known as Edward Longshanks. Edward I had defeated Llywelyn ap Gruffydd, who is known in Wales as Llywelyn Ein Llyw Olaf, the last prince of an independent Wales, at a battle in 1282. The castle was one of several formidable castles Edward embarked on to build in 1277 to control the defeated Welsh, and it was completed by 1289. The walled town attached to the castle was for Anglo-Normans only and the local indigenous Welsh remained outside the stone boundary. It was then known as Llanbadarn Gaerog and did not take up its present name of Aberystwyth until 1400.

The town remained small until the 18th century, and was compact enough to be enclosed within a stone wall which was linked to the castle of Edward I. The castle was built on an elevated promontory overlooking the harbour and the sea, with commanding views of Cardigan Bay, extending from Bardsey Island in the north to Ramsey Island in the south. The castle suffered numerous attacks and was even briefly held by Owain Glyndŵr, the Welsh folk hero who fought a fierce but unsuccessful campaign to seek independence from the English rule of Wales in the early 15th century.

In the 17th century the castle was used to store silver extracted from mines in the surrounding hills, and by the 18th century this mining activity was flourishing and the town expanded due to an increase in trade and the need to export ores.

The town became fashionable and popular with the gentry, and by 1810 boasted a significant bath house on the sea front which offered a variety of bathing experiences as well as a dining room with a well-stocked cellar. Aberystwyth grew rapidly between 1800 and 1840; the population in 1801 was 1,758, but increased to approximately 5,000 by 1835 and was 8,014 by the end of the century. It matured into a flourishing and popular resort for those who could travel there by stagecoaches which took advantage of the established turnpike road system which linked the town to Shrewsbury and beyond to the east, and Carmarthen to the south. Aberystwyth's expansion was further boosted by the coming of the railway in 1864.

Aberystwyth lies on a coastal plain with sea cliffs to the south and north, and it is supplied by two rivers whose respective sources lie deep in the surrounding hills of Pumlumon and the Elenydd. The River Rheidol has its source in the remote upland tarn of Llyn Llygad Rheidol on the south-western slopes of Pumlumon. The River Ystwyth, which gave its name to the town, flows from the damp, upland green desert in the hills surrounding the hamlets of Cwmystwyth and Pontrhydygroes.

Lead mining has a long history in the area, going back more than 4,000 years, with evidence of both Bronze Age

An early photograph of Aberystwyth prior to the construction of the pier in 1864, showing the bath house built in 1810 on the promenade

activity at Copa Hill in Cwmystwyth and subsequent Roman activity which lasted for their period of occupation from AD 43 to 410, resulting in some of the lead from the area going as far as Rome to roof temples, construct baths and to convey water. There was later significant interest in mining from the Cistercian monastic community based at Strata Florida Abbey near Pontrhydfendigaid in the 12th century. They extracted lead and silver from mines in the vicinity of the abbey.

In these surrounding hills lead and silver mining thrived in the 17th century. In 1690 a significant vein of lead was accidentally discovered at Esgair Hir in the north-east of the county by a hill shepherd working on the Gogerddan Estate near Aberystwyth. This was reported to his master and employer, Sir Carbery Pryse of Gogerddan Mansion. The baronetcy of Gogerddan had been established in 1641 and Sir Carbery Pryse was the fourth baronet of the estate, and the lineage was to continue into the 20th century until 1962 when the title became extinct. The ownership and rights to mine the ore at Esgair Hir were disputed, as the rights were claimed by the Society of Mines Royal who were directly responsible to the Crown, and who had historically maintained that the reigning monarch had the rights to all minerals. The dispute went as far as Parliament, only to be resolved in Sir Pryse's favour. In 1693 the Royal Mines Act was passed which removed the Crown's right to claim ore. Sir Humphrey Mackworth, an early industrialist and politician, together with twelve shareholding directors, formed the Company of Mine Adventurers and they acquired the mining rights and set about expanding mining activity in north Cardiganshire. Eventually they had the lease of twenty-eight mines in the area.

Lead mining activity expanded in the 18th century under the auspices of a wide range of investors, with further significant interest and development in the early years of the 19th century. At that time the mines of Cardiganshire and the boundary areas of Montgomeryshire produced around ten per cent of the total British output of lead ore, around twelve per cent of zinc, as well as contributing significantly to the production of silver.

During this period a large number of Cornishmen came to the area, together with their mining expertise. There followed a gradual decline in mining activity after that, but with a further revival in the concluding years of the 19th century. However, just a few hundred tons of ore were being produced in 1900, but some activity did continue at some locations up until the latter years of the 1920s.

This mining industry contributed significantly to the expansion of Aberystwyth, taking development well outside the old boundary walls and establishing the harbour as an important mineral exporting port. Mining was the main occupation in the hinterland surrounding the town, as well as providing related occupation in the town and harbour, which became one of the busiest ports in Wales. Lead, zinc and silver ores were transported to the town which was geographically isolated with the sea to the west and hills to the east, so maritime transport provided the solution for exporting mineral ores.

The harbour in the early years of the 18th century was only used by fishing vessels and small ships, due to a large sand bar at the entrance. Due to the increased need to export the lead ore and import goods, such as timber, limestone and culm to supply the needs of the mining industry, local agriculture and brewing, the sand bank would need to be reduced or even removed. By the end of the century work had been undertaken to partially achieve this, and there followed an increased demand for ships and shipbuilding sites around the town. Shipbuilding had been undertaken in the town since the late 18th century, and between 1770 and 1880 some 278 sailing vessels had been built in Aberystwyth.

There were attempts at dredging channels, but by the early part of the 19th century more drastic action was required. By 1836 work had commenced on constructing a stone pier to the south of the harbour entrance. To achieve this, a new road was constructed from Trefechan to the banks of the incoming River Ystwyth and a bridge constructed over the river. This enabled the construction of an embankment track with a tram road to head south to Alltwen at the far side of Tanybwlch beach, in order to transport quarried rock for the construction of the stone pier. Further work was undertaken to raise beach levels to the west of the harbour at Ro Fawr, and also the construction of groynes together with the installation of very large rocks to reduce the erosive activity of the sea. Two mooring buoys were located offshore to allow ships to moor and wait for the tide to rise sufficiently to allow passage over the bar. In 1858 the pier was extended and there were further improvements in 1870. Quays were also built on the west side of the harbour.

Shipbuilding continued, most being schooners of 100 tons or less, but some larger vessels were also built. The largest ship built was the SS *Caroline Spooner*, at 663 tons. The shipyards were located on the part of the harbour next to South Road, and became known as Shipbuilders' Row. Larger cargos of imported timber were off-loaded directly onto the beach at Ro Fawr, which is today known as South Beach in front of South Marine Terrace. Shipbuilding continued until the late 1880s, and the last ship to be built there was the *Edith Eleanor*, a 96-ton schooner, which was launched in 1881. The shipbuilding area was finally walled in during the 1890s but some timber continued to be imported up until the 1920s.

Aberystwyth and its surrounding area has a long and chequered history in which both the sea and the land have provided varied employment opportunities to those who live, both in the town and in the remote and isolated neighbourhoods of the hills and valleys of north Cardiganshire or Ceredigion. Despite the apparent isolation of these communities, there was an interdependence through the requirement for gainful employment which linked the forms of labour, industry and society, that brought together and connected families. The effects of social change through the industrialisation of employment, and improvements in transportation as well as the consequence of war, caused profound changes in rural areas where change had been limited in the past. Changes did happen and are still happening, but the land is constant and within it is the history of resourceful people who survived and prospered despite the alterations demanded of them. They are part of our past and helped to guarantee our future.

*

Into this historical context the two sides of my family, both maternal and paternal, are deeply rooted, equally in the town and the surrounding countryside.

I am fortunate that details of family history survived in written form in the handwritten notebook of my great-grandfather, Captain William James; the wartime postcards of his son, William Richard James; the detailed notes of my great-uncle, Isaac Jenkins, from Aberffrwd; and my father's handwritten manuscripts recording his childhood memories and wartime experiences. They provide a vernacular insight into the past which in turn helps explain our present and hopefully secures our future.

Bro fy mebyd is a Welsh-language term which simply means the place where one was born and raised. The term *homeland* is possibly the closest English translation, but in many ways does not convey the full sense of understanding in the Welsh context. The Welsh term *cynefin*, however, has no direct English translation. It has an understanding of being a relationship between individuals and their place of birth and the environment in which they live and to which they are naturally acclimatized. It conveys a sense of place and belonging, with a familiarity and rootedness which is temporal, physical, cultural and spiritual. It links us into a community and its shared history which is instinctive and intuitive.

Our personal cultural heritage is unique and irreplaceable, which places the responsibility of its preservation on the present generation, to be maintained and bestowed for the benefit of future generations. Our cultural heritage includes tangible culture, intangible culture and natural heritage. My family, both paternal and maternal, have a long and deep-rooted history in Aberystwyth and its surrounding area, going back for many generations. Their story has provided me with an understanding and insight into my own *cynefin*. My daughters and granddaughters will experience and explore their own respective perception of their *cynefin*, and hopefully will gain their own understanding and appreciation. Their places and experiences will be different, but will provide their rootedness, secure in the comfort of being firmly established within their own sense of place and the legacy of intangible attributes inherited from their past and securing their future.

Our family story has a long history and much of the factual evidence was documented from the early 19th century which enables the story to be told.

1

Sea

Captain William James
(1842–1917)

MY GREAT-GREAT-GRANDFATHER, CAPTAIN William James, was born in 1803 and became a sea captain on sailing ships departing from Aberystwyth harbour. He was Master and owner of the schooner SS *Margaret Evans* which had been built in Aberystwyth in 1851. He and his wife Mary Jane lived in New Street in the town. She was nearly twenty years younger than her husband and they had one son, born on 15 March 1842, also known as William James. Captain William James Senior tragically died in Nantes, France, on 19 August 1857. He was buried there, leaving his wife to bring up their fifteen-year-old son.

The young William James had already followed in his father's footsteps, and was working as a ship's boy on vessels operating out of the town, including the 88-ton schooner SS *Helena* (also built in Aberystwyth) by Thomas Jones, thus reducing the financial burden on his mother. In 1863, at the age of twenty-one, the young William James had gained his First Mate's Certificate, while serving on the schooners

Margaret Evans and *Helena*, which was testimony to his seafaring skills and knowledge. Two years later, in 1865, he had attained his full Board of Trade Master's Certificate, in Cork. He became Master of the 100-ton schooner SS *William and Mary* and a full member of the Masonic Lodge in Aberystwyth in 1862.

The Lodge in Aberystwyth had been established when the lead mines in the surrounding hills were flourishing. Miners were paid monthly on the first Friday of every month. The Mine Captains or managers would come into Aberystwyth on the Thursday to withdraw cash from the banks to pay the wages, and would spend the night at the Belle Vue Hotel on the sea front, and it was from these gatherings that the Lodge was formed. Membership was useful to the Masters and Mates of ships, as they could call upon support and assistance should they be having problems when away from home in other towns and cities.

The younger William James's intelligent and meticulous attention to detail is reflected in his handwritten notebook which records in considerable detail his maritime activity and which has fortunately survived in family records. The average active life of ships during his lifetime was less than twenty-five years, and the perilous nature of nautical life, despite many years of successful voyages and profitable employment, is reflected in his account of his share of misfortune.

Short and stocky in appearance, the bearded William James courted and married Mary Evans whose family lived on a small farm called Old Ropewalk or Ropos Fach in the shadow of Pen Dinas hill on the west side of the road from Trefechan to Penparcau. The farm looked down onto the road and beyond to the gently sloping meadows which were bounded on the east side by rows of trees and overlooking the River Rheidol as it approached the bridge at Trefechan. He moved to live at the small farm to support his wife and her elderly mother, despite having limited enthusiasm for agriculture.

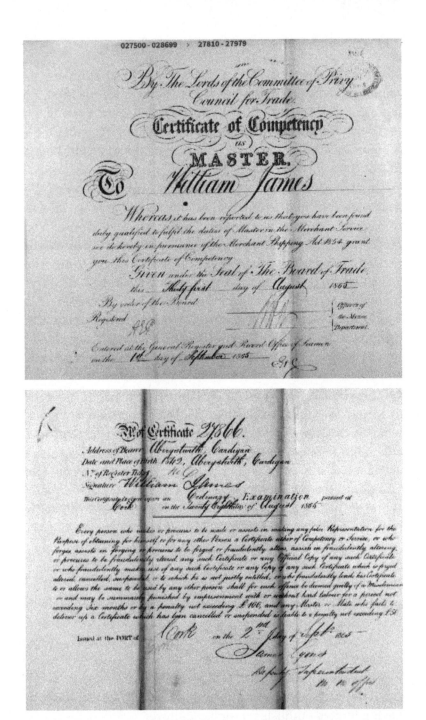

Copy of William James's Master Certificate, issued in August 1865

SS *William and Mary*

His time as Master of the *William and Mary* had been successful, with voyages around the coast of the British Isles, such as from Llanelli to Leith and London to Belfast, but also with longer and more challenging trips to Gibraltar, Bilbao and Malaga. More demanding undertakings had taken the vessel to Hamburg, via the Elbe, where there were no docks as such and the ship would be moored, tied up to posts driven into the seabed known as dolphins. The cargo was varied on outward voyages, frequently including mineral ores from the mines in the hills neighbouring the town, and would often involve shipping timber to Aberystwyth on the return, thus avoiding the need for ballast.

Voyages down the Bay of Biscay and beyond were often long and uncomfortable. Life on board ship was like living on a small island from which there was no escape. The Master and crew were isolated from the outside world and it was necessary that they get on with one another and work effectively. Harmony was vital and it was for this reason that very often the Master would choose to employ a Mate and crew who were known to him, and he would make every effort to take his selected crew with him in the event of being allocated a new vessel. Long-standing relationships developed, and this was the case with Captain William James and his ship's Mate James Latch, who was originally from Gloucestershire, who served with him for many years.

The crew would have to work hard at all hours to go aloft and trim the sails. Food at the start of a journey would be good due to fresh provisioning, but as the voyage progressed it would regress to ship's biscuits or hard tack and dry salted meat which was either beef or pork. The 1850s and 1860s were exciting years for Aberystwyth boys and those visiting the town, who would gaze at the ships in the harbour and those being built. It attracted them to a seafaring career. Life at sea was a temptation to young men from the towns and villages of the Cardigan Bay coast, as there was little in the way of employment apart from farming and mining. They would work at sea for 8s. a month. Many crews were teetotal and God-fearing men due to their strict chapel upbringing. Welsh would be the language on board. William James would make efforts to be well dressed, to set an example to the men under his command.

There had been many profitable voyages, but on occasion he and the vessel had experienced difficulties and some merchants would renege on an agreement to deliver and provide a cargo on time. William had been lucky enough to receive a good education at what was known as the British School in Skinner Street, having been taught by the Rev. William Jones and Mr John Evans, who had long experience of teaching navigation, astronomy and the theoretical skills of seamanship. William's education and experience had enabled him to deal with such difficult eventualities, and he would not hesitate to take strong action to ensure that neither he nor his crew, as well as the owners of the ship, would suffer a loss of income. His handwritten notebook records in precise detail and language how he dealt with such problems.

Demurrage was a maritime term related to vessel chartering, and referred to the period normally allowed to load and unload cargo, known as lay-time. It mentions the charges that the charterer pays to the ship owner for any delayed loading/ unloading, or failure to load or discharge a ship within the time agreed. Such a problem is recorded in his notebook in 1868.

William and Mary, *Master, 1868, Loading*

British Foreign Demurrage Association

Take note that on the days allowed for the loading of the cargo on board the vessel William and Mary *having expired, she is now and will be on demurrage for the amount of which you are and will be liable. And further take notice that should you neglect to load cargo, you and all concerned are liable to the loss and expense of my attention.*

Dated this 15th. day of May 1868

Captain William James's seafaring lifestyle meant that he was away from home for long periods, but he always ensured that his wife and her family at Old Ropewalk were looked after, and he would arrange for money to be transferred to a bank account to which she had access.

Ref. No. 6762 Alicante 15th. May 1869

For £50 sterling

At nine days date pay this to the order of Mrs. Mary James fifty pound sterling.

Value received of Captain Wm. James which place to account as advised.

To Messers Samuel Dobree & Sons

Token House Yard

London

These regular payments would be made from the many ports which he visited and provide an interesting insight into the extent of the journeys that he took.

Bank of Ireland

Ref. No. 8138 £50 15th. July 1869

Seven days after the date to pay without acceptance to the order of Mary James fifty pounds Sterling from the Governor of the Bank of Ireland to the cashier of the Bank of England London.

However, in 1870, the *William and Mary* fatally foundered off Terschelling Island off the coast of Holland. The schooner had sailed from Aberystwyth to Porthmadog to collect a cargo of slate to be delivered to Hamburg. The ship left Porthmadog on 18 April and sailed in pleasant conditions except for periods of thick fog. At around noon on Tuesday, 26 April, the Captain and crew observed a red buoy and realised they were sailing towards some dangerous sandbanks. They immediately altered course northerly, but heavily struck one of the banks. They lowered both bow and stern anchors to try to prevent the schooner sailing further onto the banks as the tide was low. The anchors held and the vessel floated freely on the flooding tide. The crew set all sails and the vessel proceeded away from the banks. The Captain ordered that the ensigns be hoisted, one with the Union Jack down and another with the Union Jack up, on the foremast, to draw the attention of the pilots to their predicament. An inverted ensign was a recognised maritime distress signal. All hands were engaged in pumping water and also throwing overboard a large quantity of the cargo in order to lighten the vessel.

At about three o'clock that afternoon, the pilot boat came alongside and Captain James requested their assistance to sail to the nearest port. However, the pilot would not venture on board as he was concerned that the vessel would not stay afloat for much longer. He begged the Captain and crew to abandon the vessel and board his, as the *William and Mary* was taking on water fast. Half an hour later the crew had no alternative but to board the pilot boat and, within a further half-hour, the schooner sank bow first and disappeared. The Captain and crew spent the night on board the pilot boat and the following morning were put ashore on Terschelling Island. They were well looked after by the local Dutch residents who were sympathetic to their predicament. Captain James wrote letters to explain their situation and they had to wait for replies before securing a passage on a steamer to London, from where they arranged a passage home to Aberystwyth. The episode had caused comment and rumour within the town as well as an account of the misfortune in the local paper, the *Cambrian News and*

Meirionethshire Standard. The press account gives a graphic account of the incident:

> *ABERYSTWYTH*
> *Loss of the schooner* William and Mary
> *This fine schooner, belonging to this port, commanded by Captain William James, son of Mrs. James of Queen Street as appears from an account given in a letter from one of the crew, left the port of Portmadoc on the 18th ult. Bound for Hamburg, with a cargo of slate, and after sailing for several days very pleasantly, the weather being fine, with the exception of a thick fog which surrounded them at times, about noon on Tuesday the 28th ult. when near some dangerous banks, they saw at a little distance off a red buoy. The vessel was immediately hauled head north-ward, and in a few seconds she struck heavily. Both anchors were put down at once, with a view to preventing the schooner from going further on the bank, as the tide was down at the time. The anchors held well, and the vessel proceeded out to sea, with the ensign flying jack down, and the jack up on the foremast to draw the attention of pilots. During this time all hands were engaged in pumping and throwing some of the cargo overboard in order to lighten the vessel. About three o'clock in the afternoon a pilot boat came alongside of them, when Captain James solicited their assistance for the purpose of attempting to steer to the nearest port, but the pilot would not step on board, saying that the vessel would not keep afloat much longer, and begged the crew to leave it and save their lives, seeing no alternative but to take the pilot's boat and the water was filling in very fast. About half-past three the crew left the schooner, and in about half an hour afterwards she disappeared, and went down head foremost. The crew remained in the boat all night in a very exhausted state, and were landed the following morning on Tershelling Island, where they were kindly treated and were expecting, when the letter was written, to get a passage across with the steamer from that place to London.*

SS *Eigen*

Despite the setback, it was not long before William James had gained new employment as Master of the barquentine, *Eigen*. Barquentine vessels had the advantage of needing smaller crews, and had a good performance before the wind and an ability to sail relatively close to the wind while carrying plenty of cargo. They were popular at the end of the 19th century. William worked hard and enjoyed his time as Master of the *Eigen*. She was built in Aberystwyth by the well-known shipbuilder John Evans, and was owned by Thomas Jones, a member of a wealthy and respected ship-owning family in the town. William's voyages were varied, as was the variety of cargo, and his travels took him to Swansea, Teignmouth and as far as Madeira and Lisbon.

Safety had improved through the introduction of what is known as the Plimsoll line, a marking on a ship's side showing the limit of legal submersion when loaded with cargo under various sea conditions, which became a legal requirement through the Merchant Navy Act of 1870.

He frequently undertook voyages which imported timber to Aberystwyth to meet the demands of the local saw mill, shipbuilding and also the mining industry.

His detailed records showed his earnings and that of his crew:

Name	Capacity	Month and days	Wages
Wm. James	Master	2 12	14 12 0
J Latch	Mate	2 4	8 10 8
J Bolton	AB	1 29	5 18 0
J Thompson	AB	1 21	4 5 0
Ed. Humphreys	Boy	1 29	2 19 0

His notes also reflect his continuing round of bad luck. He had experienced a sequence of bad weather which impacted on journey times and therefore his ability to pay his bills. He was an honest and proud man, and it grieved him to write letters requesting sympathetic consideration for his financial impecuniosities.

Gloucester, July 8th 1874
Messers. Ed Hammett Sons, Bonded Stores Merchant, Quay Parade, Swansea.

Dear Sir,

I am sorry to say that I have been very unfortunate this last round I made from your place to Madeira and thence Casablanca.

We encountered very rough weather which carried away boat sails, bulwarks and everything moveable. The vessel has been transferred entirely to my name. I had two heavy club sails to pay and to reinsure for the new gear. My insured freight was not sufficient to pay the expenses. The vessel is now ready to sail for Limerick, from which place I hope that I will be able to remit the money.

I hope and trust that you will be able to make it convenient to wait a little. I will sure to remit the money as soon as possible with interest on the same as the time is getting long.

I remain your obedient servant
William James
Master of the Eigen

His profitable voyages were accurately and diligently recorded in his records.

These profitable voyages enabled him to save some money, and he had already agreed the terms for the purchase of the vessel and become its primary owner, in the hope that others would invest as part shareholders, as was the custom in Aberystwyth. His notebook provides evidence of his venture, and a copy of the agreement he had succeeded in negotiating to pay for the ship in three instalments.

Aberystwyth, 21st April 1875
Received from Capt. William James the sum of three hundred pounds deposit as per agreement on this same date with myself for the purchase of the Brigantine Eigen *of this port and being part payment of the whole of the purchase money agreed to be paid, namely eleven hundred pounds and which said sum of three hundred pounds money I have endorsed on the back of the agreement.*

Thomas Jones
21/4/75
£300 £300   £250   £250

1876	Limerick	£	s	d
June	Ship Brokers Bill inwards	4	"	11
	do outwards	2	6	6
	Labours discharging	5	"	"
	To Allowance do	"	5	"
	To Labour loading Timber	5	"	"
	To repairing galley stove lamps	1	"	"
	To Ship Chandlers	1	10	7
	To Pilotage down River	1	13	"
	To Damage done to Hooker	1	"	"
	To Grocers Bill	2	12	1
	To 4 cwt Potatoes & Vegetables	1	5	"
	To Butchers Bill	2	9	1½
	To Night Watchman	1	"	"

1875	Mary port	£	s	d
April 23	To Hobler outwards Aberystwyth	1	"	"
April 28	To Boat assisting Mooring Ship	"	2	"
April 29	To 3 Men discharging Ballast 55 tons 1½	1	"	7½
May 3	To Trimming Cargo 220 tons how 1½	1	7	6
do	To Ship Brokers including all Charges	9	14	8½
May 4	To Boat assisting outwards	"	2	"
	To Gatemen's gratuity	"	2	"
May 3	To Grocer's Bill	2	2	"
	To 7 lbs fresh Meat 7 lbs oct & Vegetables	"	5	"
	To New paper Box	"	"	4
	£	15	16	2

Example of Captain William James's notes in his own hand

Copy of purchase agreement in Captain James's notebook

However, his record of misfortune continued and this elegant vessel was wrecked in the far distant port of Casablanca. His notes include the draft of his letter to the Mutual Ship Insurance Society who had established an office in Aberystwyth in 1853 and had provided considerable service to the seafaring community in the town:

Casablanca, 18th Dec. 1876
Mutual Ship Insurance Society, Aberystwyth.
Gentlemen,
 Not seeing it necessary to write the following day owing to no mail leaving here. A survey having been called after the discharging of cargo. I regret to inform you that the vessel has been condemned a total wreck. The tide ebbs and flows in the vessel, her bottom is breaking up and washing ashore. The keel is done and her stern and bow is all open. We have dismantled the vessel of all her sails, ropes and most of her spares. Vessel laying broadside on the sea having a heaving list. Left standing mast and rigging to protect the masts. All will be sold tomorrow by public auction before the vessel will break up

as it is very stormy weather here daily. There are several vessels out in
the bay and most of them have lost an anchor. The Kate *of Brixham*
the day following her arrival drifted a long way. Should the weather
have continued a little longer she would have followed the Eigen.

By the first mail the copy of Protest and other documents will be
sent to the Board of Trade through the British Delegation of Tangier.
A copy of these I shall make myself.

Gentlemen, I remain your obedient servant
William James

He had been fortunate that, once the disposal of the remains
of the wreck had been completed, he met Captain Benjamin
George of 25 Millicent Street, Cardiff, who was at the port in
his vessel, the brigantine *Syren*. Captain George was able to
take him and his crew back to Cardiff, from where he was able
to find his way back to Aberystwyth.

His time at home after this misfortune weighed heavily on
him, and he was both relieved and delighted when, at last, he
was able to go back to sea once more. His long-established
connections in the town and further afield bore fruit, and his
membership of the Masonic Lodge in Aberystwyth worked
in his favour as it had done in the past with some of his
misfortunes. He had written letters requesting a vessel, and his
persistence had paid off:

To Messers Phillips and Rees
Ships Brokers
Swansea
Dear Sirs

I beg to inform you that I am now at home after the misfortune I
had at Casablanca of losing the Eigen *(Brigantine).*

Should you know of any vacancy I should be very glad to accept
the same and I should be very much obliged for your kindness. It will
take some time before I get the insurance and there is nothing here at
present.

I remain yours faithfully
William James

As a result he became Master of the schooner *Clarissa*. She had been built in 1855 in Aberystwyth and had sailed previously with Captain James Watkins as her Master.

The records from his notebook outline many voyages, including a two-week voyage to Cadiz in south-western Spain in 1879. The log entries provide a vivid insight into the busy life of the crew when responding to changing weather and setting appropriate sail to safely undertake the voyage. From the log entries it is possible to accurately plot the journey and gain an impression of the variable weather. The following extracts indicate conditions at the start and the end of the voyage:

Clarissa, *25th April 1879*
Strong Westerly wind. Took in mainsail and settled the staysail.
Lat 49-5 Long 5-50W

7 May
Fair with variable light wind. Tack ship WSW. Chipiona light house SES.
At 9 fresh southerly wind. At noon Pilot came on board. At 2 in Cadiz Bay in fair weather with 30 fathoms of chain. End of voyage.

8 May
Fair Westerly breeze At 5.30 a.m. Pilot came on board. I weighed and made sail for the Arsenal. At 8 came to anchor at the discharging berth.

The vessel returned to Aberystwyth with another cargo and his notes record yet another turbulent episode involving one of his vessels.

The *Clarissa* had been moored on the north side of Aberystwyth harbour in an area known as St David's Wharf at Pen-yr-angor, and was lying hard against the steep stone walls.

The ship had been loaded with galena – lead ore, which had been brought down to the harbour from the hills to the north of Aberystwyth. The weight of the mineral ore reduced the volume of cargo which could be carried but, nevertheless, provided excellent ballast for the journey ahead. Leaving the harbour would always be easier than entering, as the ship would move gently downstream at high tide and use the flow of the current of the River Rheidol to move forward smoothly into midstream until the adjacent flow of the River Ystwyth would cause the bow to turn and line the vessel up parallel to the stone jetty and out over the infamous shingle bar to open water. There would be no need to call on the assistance of the hobblers waiting on the jetty.

The reassuring creaking movement of the ship's timbers and the flapping of billowing sails would have been a relaxing reassurance that another journey was underway. The crew of four men and a boy were experienced local people, some of whom had been with the ship for more than two years and served with her previous Master, Captain James Watkins, and knew her well. The Mate and another member had served on many occasions with William James. They worked as an efficient team to maximise the sail required to allow steady progress. The brisk prevailing south-westerly breeze would ensure that good advancement would be made north up Cardigan Bay towards the tip of the Lleyn Peninsula and around Anglesey to the port of Liverpool. They knew that this short passage would be followed by a longer journey and that they would be away from home for some days.

The *Clarissa* was not a large vessel but had the capacity to carry a worthwhile cargo. She had been built by John Evans in Aberystwyth, and was a schooner rigged with a standing bowsprit and two masts. Her single flat deck and square stern allowed easy movement to work the sails, and her elegant figurehead emphasised her progress through the swells of the Irish Sea. The journey north passed without incident and they entered the River Mersey on a flooding tide which assisted progress from the sea for just short of three miles to the port of Liverpool, enabling the vessel to proceed upriver despite the river's current. They slowly progressed to one of the many docks and secured the vessel.

William James went ashore to arrange the removal of their heavy cargo, and a band of burly dockers would unload. His crew remained in the vicinity of the vessel to oversee the unloading, and then to ensure that the hold was thoroughly scrubbed out in readiness for the next cargo which their Master and First Mate would need to negotiate.

Neither Captain nor crew would venture far from their mooring, as Liverpool had grown into a cosmopolitan place of some size and had been granted city status earlier that year, in 1880. However, with the expansion of the city there was also greater incidence of crime and thuggery, and the narrow streets surrounding the dock areas were the haunts of gangs of thugs who loitered outside the numerous public houses. They would intimidate any passerby or those entering the premises and demand money to buy drinks. The gangs of thugs were very territorial and enjoyed fighting and some, like the High Rip Gang, would target dockers and follow them on their pay days to extort money. The streets were not the place for the crew of the *Clarissa* and they looked forward to the successful completion of negotiating a fresh cargo so that they could return to the comparative safety of the sea.

As a frequent visitor to the port, William James was familiar with many who worked there and a bond of trust grew despite the prevalence of criminality in the city. He had visited the offices of the shipping company Heidfeld in Old Hall Street and then returned to his vessel to inform his crew that he had negotiated a cargo of corn to be taken to the port of Avilés in Spain.

The corn arrived at portside in sacks and, to save time, it was decided to commence loading the sacks rather than taking the trouble to install shifting boards and loose-loading the grain into the hold. He and all his crew went ashore to meet with the shipping officer and to formalise the necessary documentation for the journey and the discharge of the cargo to the appointed consignee in Spain. He left the shipping agents, crew and tallyman to load the sacks into the hold. Final checks were made on water and fresh supplies, and the crew prepared the vessel for an early morning departure on the ebb tide to assist progress down the Mersey and out to sea. The weather was settled for the time of year and the impending voyage looked to be straightforward.

In the early hours the vessel progressed down the river passing the bustle of the busy docks and the varied array of vessels moored there. They sailed west out to sea and out beyond Anglesey to head south.

Captain William James had ensured that his accounts were accurate and the overheads of the visit to the port were recorded.

	£	s	d
Port of Liverpool			
Medicine for medicine chest	1	6	6
Carpenters' bill	10	7	6
Repairs		12	1
T H Williams, ship's chandlers	13	5	11
Loading of Corn	2	17	11
Interest on insurance paid	1	12	5
Kelly the Butcher's bill:			
beef 1 cwt., 18lbs potatoes,			
112 turnips, vegetables	4	10	0

South-westerly winds assisted progress and life on board settled into a pattern of almost monotonous routine activity. Captain James would spend time in his cabin at the stern of the ship and in close proximity to the First Mate's accommodation. The crew and the boy were forward in the bow section, where their bunks provided respite from the arduous duties of life on deck, despite the damp and cramped conditions. Their beds were simple narrow bunks on which there were mattresses of sacking and straw which would remain damp after a period of inclement weather. Between the Captain's and crew's accommodation lay the cargo holds. The galley was situated forward on the deck, just behind the sail locker, and one of the crew took responsibility to prepare their simple meals. They would often be cold, wet and hungry and, as the voyage progressed, they would resort to the 'pound and pint' diet where meat and ship's biscuits were allocated by weight, and the dried peas, flour and oatmeal measured in pints. They would eat cracker hash or sea pie, which consisted of layers of salt beef, peas and powdered ship's biscuits washed down by small quantities of precious water.

Progress down the Bay of Biscay was without incident and within days the vessel had arrived at Avilés on the northern coast of Spain, and west of the better-known port of Bilbao. The *Clarissa* sailed gently south in the wide protected estuary under the watchful eye of the pilot, and arrived at the port with the crew knowing that they were in a secure and sheltered location. On the far side an isolated ship was moored with a yellow flag flying on her mast, indicating that she was in quarantine having arrived from a distant port with a known problem of illness or disease. That ship would remain there for several days until the crew and their quarters were smoked or fumigated to avoid any transmission or infection. Nothing would be allowed on shore, but food would be allowed to be taken on board.

As a Master of vessels, William James had gained considerable experience. As noted, having gained his First Mate certificate in Aberystwyth in March 1863 at the age of twenty-one, he had gone on to qualify as a Master Mariner two years later in Cork. His voyages on several different ships had taken him to a wide range of ports and countries, from Leith in Scotland to Casablanca in Morocco. He was an intelligent man and, through his responsibilities, he had gained a basic working knowledge of the languages of Spain and France.

Captain James went ashore to contact the consignee for the

cargo and soon his representative arrived to inspect the cargo and agree for its discharge. However, it was necessary to wait for Custom House officers to arrive before unloading could commence. So both he and the crew spent a short time ashore, having left the First Mate, James Latch, aboard to secure the vessel.

When they returned the discharge of the cargo was in progress and the grain was being transferred from the onboard sacks to the sacks provided by the local merchant. These sacks were approximately half the size of the cargo bags and the work was well underway. However, it became apparent that the Custom House officers who were tallying the discharge were agitated and the process begun to slow down. Captain James strode the quay and, realising that the work had stopped, he was then summoned to speak with the tallying Custom House officers. They were concerned that there appeared to be a discrepancy between the original cargo manifest and the quantity of grain unloaded. The Custom House officers were adamant that there was an inconsistency. They had no other course of action but to impound the vessel until a satisfactory explanation could be found. Failing that, the vessel would be subject to a fine of £750 which would need to be paid prior to its release from the port. This sum was equivalent to over £80,000 in modern equivalence.

Despite his basic linguistic skills, William James was finding it difficult to argue and reason in a language unfamiliar to him, and it rapidly became evident that their stay in Avilés was going to be prolonged. He gathered his crew on deck and explained the situation to them, but assured them that he would immediately be working on resolving the problem. Meanwhile, they were to get on with their routine duties and prepare the ship in readiness for another cargo and an onward voyage. He went ashore and visited the merchant's offices to request some help, and was pleased with the assurances received from his Spanish contact. Captain James returned to the ship and retired to his stern cabin, and set about writing a series of urgent letters.

The vessel's owner was Edward Jones of Aberystwyth, a wealthy family man who had a long association with shipbuilding in the town. Edward Jones's father, Thomas Jones, owned several ships as well as some thirty houses in Aberystwyth. As noted, Thomas Jones had had the ship built in 1855, and had passed on the ownership of his vessels to his son Edward in June 1879. Edward Jones would be anxious at the loss of income from the impounding of one of his vessels.

The notebook contains the draft of the letters Captain James wrote:

Ed. Jones Esq.

Dear Sir

I really don't know what to write, only the time is going on and my duty is to write. I am quite miserable over this unexpected affair. Thinking that I have been all my lifetime trading to Spain and never had any trouble with Customs before. I wrote to the Shippers Mr Oscar Heidfeld, 41 Old Hall Street, Liverpool, stating to him how the vessel is situated by the Custom House authorities here, owing to the invoices they made out these Bills of Loading and got me to sign for 232 sacks and only 207 sacks were found in the hold here. There were two Custom House officers on board tallying the sacks as they were emptied out to measure the corn into the merchant's sacks which were about half the size of our sacks. They found the quantity of corn which the 25 sacks would contain in bulk and not in sacks. As they were in a dreadful hurry in loading they must have let loose the contents of the 25 sacks into the hold and returned the sacks on shore and marked the ticket as bulkhead sacks. If the 232 sacks were put on board there would be considerable more corn turned out. I beg of him should you find the 25 sacks to acknowledge the same by writing either to the merchant here or to me. What seems such a ridiculous thing, is to find the vessel fined for £750 for 25 empty sacks which was given on loan to this vessel as a substitute for shifting boards. I regret now daily that we did not put up shifting boards instead, but it is too late now. I thought I was doing the most economical way. My Merchant here is trying his best. He wrote

*to the Member of Parliament which is a good friend of his. I saw a
reply from him today saying that the report of it has not come from
the Custom House here yet and, as soon as it comes, he will start to
look into it to release the vessel. Trusting that we will get free as soon
as the weather changes. It is very rough here today. They persuade
me that it will come to nothing, only it will require a little time to
correspond with Madrid. If we had been free, would still be here for it
is frightful rough weather daily.*

He wrote another letter to the merchant in Liverpool whose
cargo he had transported and which was now in question.
He went ashore to arrange for these letters to be sent, and to
negotiate further help from the Spanish authorities.

Avilés, 17/2/80
To Oscar Heidfeld Esq., Old Hall Street, Liverpool
Sir,
*The Consignees are well satisfied with the cargo of maize discharged
to them from the Clarissa. But I am extremely sorry to say that I have
got into a fearful mess with the Custom House authorities, owing to
25 empty sacks being short of manifest. We could only find 207 sacks
instead of 232 as stated in Bills of Loading. If 232 sacks were put on
board at Liverpool, the same would be seen discharging here. The
contents of them is found in bulk, not in sacks. Should the 25 sacks
missing not be put on board there would be considerably more cargo
missed out. The 25 sacks must have been let loose in the hold and
returned on shore. I entirely trusted to your Tallyman, more so than
any Mate, as we were all absent from the vessel while the loading
was going on for about an hour while with the Shipping crew and
Shipping Officer. The Customs here was not satisfied by the way
the Manifest and Bills of Loading were made out. Nothing worded
in either about the sacks to be returned by the vessel. I took it from
you verbally to look after the sacks and return them to England. So
I have kept the sacks on board and I can assure you that I was quite
astonished when we could find only 207 sacks instead of 232. There
were two Custom House officers then boarding during the whole
time of discharging.*
    Dear Sir it seems such a ridiculous thing to fine the vessel for

£750 for 25 empty sacks. Should you find out the 25 sacks, I would be extremely obliged to you to acknowledge the same by writing to me or to the Consignee, for they have tried their endeavours to get me out of the bother I am in, through such a trifling matter as 25 empty sacks. Trusting that you will do your best to get me out of this trouble.

I remain Dear Sir
Yours faithfully
William James
Master of Clarissa

Days passed and there was little news except for the fact that the Custom House officers were adamant that there was a discrepancy. William James's crew had done all he'd asked of them, and the vessel was in excellent order. Time had been spent on routine maintenance work and there was only so much time they could spend on holystoning the deck. The port was busy with other vessels coming and going; the ship which was quarantined had lowered the yellow pennant, indicating that the smoking and health screening work was complete. It had then left with a new cargo.

Captain William James was becoming increasingly anxious that the cost of his enforced stay was having a serious impact on the revenue earned from the delivered cargo which had gone to the consignee. He had made numerous contacts, both in Avilés and at the adjoining port of Gijón which housed a local office of the British Consulate in the province of Asturias. These had enabled him to communicate his predicament to London and to the capital in Madrid. He thought that he might need to release his crew and enable them to find employment on another vessel which would be returning home to Wales. However, they remained loyal and spent some of their time assisting with small amounts of work at the port in the hope that they too would soon be on their way. Time for them passed slowly and passed anxiously for Captain James. Further letters were received and he sent a telegram:

Telegram to owner, from Avilés

Jones. Clarissa. *Aberystwyth.*

Letter received. Cannot sail until fine guaranteed. What has been done? Without, vessel must remain.

William James

p.s. Schooner Thorburn of Fleetwood *here.*

Damaged 241 sacks of Indian corn and wheat.

The same had to be lightened out to sea.

He wrote another letter to the owner in Aberystwyth, as well as corresponding with his merchant in Liverpool.

Avilés 28/2 /80

Dear Sir,

I received a letter today from Messers Jones Price and Co. of London, dated 23, acknowledging the receipt of my letter and likewise H.M. Consulate forwarded to them from you, and also a letter from our Consignees here, and that they have been with Novelli & Co. of London which is our merchant. That they knew all about the Clarissa's *case and they have written to the Spanish Ambassador at London so that they will consult with Madrid and get the vessel released soon. But nothing mentioned about the guarantees as the Vice Consul requested to give Novelli & Co. the guarantee. So the Consignee here is quite willing to give the guarantee if Novelli & Co. would authorize them to do so. H.M. British Consulate being at Avilés today on business concerning other vessels, I had an opportunity to speak with him and gave him the letter I received. He informed me that the Report of Defence has been going to and fro between the Custom House and Gijón about the guarantee and unless they will get it, the demands will not be sent to Madrid. H.M. British Consulate advised me to send you a telegram, which I did to save time, as it takes such a long time to correspond. In the meantime the Custom House has stopped the vessel as security for the fine. If they had done so at first there would have been a deal of time saved. H.M. British Consul hopes that by the time he returns to Gijón tomorrow, that he will find that the documents will be on the way to Madrid. Once they get there he is almost sure that the vessel will be soon released free.*

Wishing that it will be the case
Dear Sir I remain
Your obedient servant
William James

There seemed little progress however, and days passed into weeks with the crew now becoming restless as they could see the regular arrivals and departures in the port, and that their Master was becoming increasingly irritated by his circumstances and inability to resolve it. This was made worse by the fact that they knew that he was innocent and that there had to be a simple explanation for the problem, but the Custom House officers were unrelenting in their interpretation of the inconsistency with the cargo of corn. The crew remained loyal to William James and, being a small team, they depended on each other. There were crews operating on ships who would undertake a voyage and, on receipt of their wages at the port of arrival, would leave and cause the Master to look for a fresh crew to undertake the voyage home. Men such as these were known as runners.

William James received letters on a regular basis from Wales, London and Madrid, and by April, after a period of over eight weeks, there was finally some hope that the issue could soon be resolved. He wrote again to the ship owner:

Avilés, 9/4/80, Ed. J. Jones
Dear Sir,
I am in receipt of your valued favour of the 30th. Instant and likewise of Messers. Jones Price & Co. of the same date. Glad to learn that you are both doing your best towards getting the vessel released. The only favourable news I got to say is that I showed the second letter with my Merchants and from their friend the Member of Parliament for Avilés saying that he had seen the Report of Defence of the British Schooner Clarissa*. He was astonished at such a fine. He said that the fine will not be ticketed, that the vessel will soon be free. I also enclose to you a letter received from our Mr. Price's counsel, Gijón.*

Hoping that I will have better news soon
Your obedient servant
William James

Copy H.M. British Consulate, Gijón
William Penlington
Captain William James
Dear Sir,
A few days ago received a letter from my Consul General saying that
your case has been brought before the British Minister at Madrid and
that he expects soon to hear a satisfactory settlement. It is likely that
you will hear of this settlement at Avilés before I can here, in which
case drop me a line.
P.S. I have heard nothing from your accusers.

William James was delighted to receive further communication that, finally, progress had been made and that his enforced stay in Avilés would soon be over.

Copy
British Vice Consulate, Gijón, 10/4/80
I have to inform you that the British Consul at Coruna under date of
the 5th. instant informs one that his Excellency the British Minister
at Madrid has reason to believe that the fine on your vessel will be
remitted. The Spanish Custom Authorities having decided that the
grain shipped in sacks shall be considered as grain shipped in bulk.
Therefore I have to request you to inform me at the earliest moment
if any order to the above effect has been received from Madrid by the
Custom House at Avilés, in order that I may communicate your reply
to his Excellency the British Minister. The Collector of Customs at
Gijón is of the opinion that the order for your release will go direct to
Avilés without passing through this Custom House.

William James wrote further letters to the ship's owner and prepared the text for telegrams to be sent outlining the fact that he hoped soon to be underway.

Copy of Telegram
Jones. Aberystwyth. Clarissa Released.
Bilbao charter cancelled. Re-charter load Avilés
England eight shillings. Bristol Channel Nine Liverpool
Letter to follow
James

Ed. J Jones
I enclose you this letter of our Vice Counsel which I received this
morning. I am also very glad to inform you that my Merchants
received a telegram this morning from their friend the Member of
Parliament for Avilés saying that the vessel is released from this
unpleasant bondage. That the order will be sent to the Customs
House here in a few days. Meantime I have ordered ballast. Hoping
that we will be released soon and ready to sail next Monday.
Dear Sir, I remain your obedient servant
William James

However, things did not proceed as well as he had expected
as there were the inevitable delays in seeking total confirmation
of the outcome and written proof to be provided to the Custom
House officers at Avilés. This was resolved and, by 18 April,
he was able to communicate that the vessel would finally be
underway.

Copy Letter, Avilés, 18th. April 1880
E.J. Jones Esq.
Dear Sir,
I sent you a telegram on the 16th. instant saying that the vessel had
been released free of the enormous penalty. The same day a cargo of
mineral turned out to load here. I was sorry then that we had taken
on ballast. Still I thought it much better to load here than making a
passage to Bilbao.
Bilbao has a very bad bar and a great many steamers trade there. We
might be kept very long there. We have commenced taking a cargo
this morning.
Dear Sir, I remain your obedient servant
W. James

William James's luck did not change easily; the weather deteriorated and he had to wait another day before leaving. He sent yet another letter to both inform and appease the ship's owner.

Avilés, 19th April 1880
I am sorry to inform you that we are still here laying wind-bound, loaded and bound for Cardiff. The equinoxial gales have settled in. It's fearful rough weather. I hope it will moderate before the tides get low. There is difficulty in getting out of this place without fair wind and fair weather. It's very narrow and shallow.

I am longing to get out of it to make up for lost time. It's one of the most expensive places that I was ever in.

Finally, the *Clarissa* set sail with accompanying pilotage and progressed out to sea to sail briskly up the Bay of Biscay. William James wrote in his notebook that they 'had done four hundred miles since leaving port with wind in our favour and a good breeze'.

His notes also recorded that the episode had been expensive:

Disbursements at Avilés	£	s	d
Pilotage & boats from sea to port	8	5	4
Discharge labour	1	8	0
Brokers' bill	23	2	7
Ballast	1	8	0
Discharging Ballast & carting it away	1	2	0
Pilotage from Avilés to sea	4	4	0
Ship's chandlers' bill	5	3	
1 cwt. of biscuits	13	0	
3 telegrams	12	0	
Postage	10	0	
1½ cwt. potatoes	6	0	
Vegetables	2	5	
186lbs beef	5	5	8
5 cwt. coals	6	0	
Total	47	10	3

The *Clarissa* delivered her cargo to Cardiff and then took on a further cargo of timber to be delivered to Aberystwyth. Sailing north up Cardigan Bay, both the captain and his loyal crew would have been glad to see the outline of the stone jetty on the horizon, and knew by the state of the tide that they would not need to stay off-shore for long to wait for the black ball to rise on the signal at the end of the jetty which indicated that there was sufficient water over the sand and shingle bar to permit safe entry into the harbour. They would sail slowly into the familiar surroundings and work to ensure that the twin towers of the old chimneys on Pen-yr-angor were aligned to guide them safely in, and were happy to accept the assistance of hobblers to turn the vessel upstream and into a mooring.

*

William James's records show that he continued to work successfully on the *Clarissa* for a number of years after this, with voyages ranging from Aberystwyth to Lisbon, Nantes and many other ports, and the transportation of a variety of cargos. However, the vessel met an unfortunate end in July 1886 whilst bound from London to Douglas on the Isle of Man with a cargo of cement. The ship, under the command of Captain John Jones of High Street, Aberystwyth, had experienced several days of bad weather and had started leaking on a Friday in the proximity of the Cardigan Bay lightship. The crew of four, which contained two of John Jones's sons, Evan and David, took to their lifeboat from which they were rescued by the crew of the steamship *Mersey* of Liverpool. The *Mersey*, with her crew of ten, then proceeded her way to Liverpool where she was bound from Newport with a cargo of coal for the African Mail Steamship Company. All went well until the Saturday morning, but at two o'clock the vessel struck a reef of rocks, called Penrhos Point, near South Stack, in fog and drizzling rain. The four men from the *Clarissa* once more took to the lifeboat with some of their hastily gathered possessions, and made for Holyhead which they reached with difficulty. Information on the disaster was conveyed to the coastguard, who immediately set off to rescue the crew of the *Mersey*. However, they had left the stricken steamer with some of their clothes and bags of possessions. The crews of both vessels were well looked after by Captain O.H. Parry of the Shipwrecked Mariners' Society, who assisted them in reporting the incidents and arranging passage home. The double disaster befalling the crew of *Clarissa* exemplified the perilous nature of seafaring as a means of earning a livelihood.

*

The railway from Shrewsbury arriving in Aberystwyth in 1864, and a line from Carmarthen arriving in 1867, had a severe impact upon maritime trade in the town. It cut profits for Masters and ship owners. Goods continued to be imported into the town, especially timber, but very often there were no cargos to be loaded and exported. Exports from the port declined from 2,385 tons in 1880 to 700 tons by 1890. The lead mining industry had gone into decline as demand for ore had dropped to such an extent that in 1884 it was fetching a price of £7 a ton, whereas in 1873 it had been £15 a ton. Ore was now imported from Spain, America and even Australia. The railway also put an end to the animal drovers and their travelling way of life, as now cattle and sheep could go to Smithfield in London in less than a day using the railway.

Captain James's livelihood took a turn for the worse in the close maritime community within which he functioned. It became increasingly evident to him that escape from his circumstances could be difficult, as the loss of a ship was not an isolated occurrence in his experience as a master mariner.

Times had been turbulent at a national level, and there

had been much talk and concern that an election might not improve the situation in the country. William Gladstone was prime minister for the third time, with a Liberal Government, and his support for Home Rule in Ireland had resulted in riots in Belfast. The Conservative Party won the election, but depended on the support of the Liberal Party in order to govern, increasing the uncertainty at national level, which filtered down even to west Wales.

In the late 19th century Sundays in Wales, and especially in Aberystwyth and Cardiganshire, would be transformed, as a hush would befall the county and the township. A hush so deep that it could be felt even in the surrounding countryside surrounding Old Ropewalk farm, the cowsheds, hen coops and sheepfolds. The people went to their chosen places of worship. William James attended the Tabernacle Methodist chapel in Powell Street, Aberystwyth. Tabernacle had been established in 1785, and grew to become an important centre for Welsh Methodism. It was also known as Capel y Groes, and required numerous episodes of rebuilding. By 1878 it had been enlarged to a size that could accommodate 1,500 worshippers. The opening ceremony of the new chapel was an important event in the town, and celebrations lasted from a Tuesday until the following Sunday, over which time nineteen sermons had been preached and eleven visiting Methodist ministers had taken part.

The chapel had a large commanding frontage, and the congregation entered its imposing interior, with its large curved U-shaped gallery of varnished pitch pine seating, supported by elegantly fluted iron pillars, to take up their seats in the rows in front of the ornate section of pews reserved for deacons. Services would be long, with an extensive sermon delivered with vigour from the pulpit where the minister, the Rev. Thomas Levi, could look down upon and observe the entirety of his faithful congregation. After the services the congregations would depart, and there would be considerable social exchanges outside the wrought-iron entrance gates. This provided opportunities for the town's residents to meet and exchange news. Amongst this gathering would have been many of the partial or total ship owners and those with an interest in vessels operating from Aberystwyth.

Prior to the Sunday Closing Act of 1881, many of the men would have walked to the Unicorn tavern at the corner of Pier Street where they would drink ale and continue discussions in relation to business and commerce in the town. However, this custom discontinued after the Sunday Closing Act. Many Methodists had shrewd business interests in shipping, and there was a distinct contrast between the merchant classes and the labouring poor in the town. It would have been at these gatherings that William James would have met many who had significant interest in the declining maritime trade operating from Aberystwyth.

Tabernacle chapel continued to function into the 21st century, but finally closed its doors in 2002. Sadly, on 4 July 2008, it was ravaged by fire, and was demolished on 11 July 2008 having served the community for over 125 years.

William James's notes indicate that he had decided that pride would need to take second place over the necessity of finding another ship so that he could escape the claustrophobic confines of his small farm. He would write to ship owners outside Aberystwyth in the hope that he would be given command of a vessel once more.

The draft of letters, in his fine elegant handwriting in black ink, reflect his frustration at not being at sea, and he contacted many of his associates in the hope of gaining another vessel.

Dear Mr Byford

I beg to inform you I have had enough of living in the same house as my mother-in-law. As you know that, on the best of times, she is not much company on account that she is old and very peevish.

Should you know of any vacancy for a Master, I should be very glad to accept the service and I should be very much obliged for your kindness.

You know where to find me. I hope that you would excuse me for taking the liberty of writing to you on the above subject.

I remain yours faithfully
William James

He wrote other more tempered letters in the vain hope of success at a period when maritime activity in the town was in decline.

To Mr John Purivell
Dear Sir
I beg to inform you that I am now out of employ, and should you know of any vacancy I should be very pleased to accept the same and I should be very much obliged for your kindness.
I remain yours faithfully
William James

To Mr John Curwen
I beg to inform you that I am now out of employ, and should you know of any vacancy I should be very glad to accept the same and should be very much obliged for your kindness.
To acquaint you that I am caught at last in the bounds of matrimony.
I remain yours faithfully
William James

There are no further records of maritime activity within his notebook, and it becomes evident that William James had to reconcile himself that there was no profitable future at sea and that he would finally have to accept that he needed to put his energies into the family farm which he had married into.

The farm was located where there was access to both sides of the rivers entering the harbour, and it was also within reasonably close proximity to Trefechan.

This area, south of the River Rheidol, is as old if not older than Aberystwyth. Due to its closeness to the harbour, it became a form of industrial suburb to the town through the presence of shipbuilding yards, limekilns, timber yards, breweries, brickworks, rope walks, and smithies. This small neighbourhood had several taverns as well as four lodging houses which could house vagrants seeking casual employment. The area's residents were significantly poorer than those in the town beyond the bridge. There was often trouble in the vicinity, fuelled by alcohol and poverty, in some of the narrow side streets and cottages.

William James decided that there was money to be made in milk, so he concentrated his efforts on expanding the small herd that they had. His entrepreneurial instincts from his life at sea inspired him to look at other opportunities for the small-holding too. He began to read up on gardening and growing crops which could be sold locally. He sent for seed catalogues from Suttons Seed Company, which by then was well established in the country and had received Royal Patronage from Queen Victoria in 1858.

He also invested in a substantial cart and a pair of strong horses to pull it. It would be used to deliver milk, but his opportunistic nature also prompted him to use it to supplement his income by offering his services to deliver goods and materials in the town. As ever, he kept meticulous records of his income and outlay which had been part of his life as a ship's Master.

Aberystwyth Town Corporation	£	s	d
January 1 Carting of gravel from the beach			
to Corporation ground near Plascrug		3	9
January 3		7	6
January 4		7	6
January 5		7	6
January 6		7	6
January 7 Carting of gravel to Skinner Street		7	6

January 8		7	6
January 10 Carting earth from Northgate to			
Queens Road		7	6
January 11		7	6
January 12 Carting from Chalybeate Street to			
Queens Road		7	6
January 13		7	6
January 14 Carting from beach to Southgate		7	6
Total	4	6	3

His new life now meant that he was permanently at home on the farm and it enabled him to spend more time on one of his interests. He was a talented artist and would sketch scenes at his beloved harbour and also of the locality. He would then meticulously paint water-colour pictures from the drawings made, and enjoyed depicting the sailing ships which had been such a big part of his life. He was impressed by the work of Alfred Worthington who had come to the town in 1870 to work as a photographer, and who also worked as a self-taught artist. Worthington painted hundreds of paintings for both local and tourist markets, as well as decorating domestic items such as fire-screens. William James was an admirer of his work, and his own paintings – some of which are to be found in the photograph section – had a distinct similarity to those of Worthington and provided him with an opportunity to relax in his hard and busy life.

Life on the farm was difficult, and it took its toll on his wife whose health deteriorated. William James now worked the farm alone, and did his best to get as much help as he could. She passed away on 19 October 1892, aged fifty-seven.

He inherited the property and continued to farm. He took up with Jane Williams from Tanllan Mill, Llanfihangel-y-Creuddyn, whom he had employed as a housekeeper to assist with domestic chores of the farm. She was a slight woman, and thirty years his junior. She lacked the robustness of his first wife Mary, but she had a serious and meticulous nature which suited

Captain William James at Old Ropewalk Farm

him well for maintaining the household and his fastidious record-keeping. In 1893 they married, and on 15 August 1895 their son William Richard James was born. It was a difficult birth for Jane James as she was a small woman, but she recovered well and the family prospered.

William Richard, or Wil as he was known, was of slight and wiry build, but grew up to be a strong and healthy boy. He was educated at the school in Trefechan and attended Sunday school. As he matured he helped his father on the farm and played a vital role in the success of the dairy business.

The town and the surrounding area were prospering due to the arrival of the railway. The pier had been refurbished and restored, and opened by the Princess of Wales in 1896, the year after William Richard's birth. There was plenty of work and therefore plenty of people requiring milk and other services. The beach area around what was known as Ro Fawr at the south end of the town was being developed. There was plenty of profitable work to be had. In 1902 the promenade was extended south from the pier, around what is now the Old College and round the base of the castle toward Ro Fawr and the harbour. Rubble was transported to the worksite along the sea front by horse and cart. The promenade extension cost £16,000, but was considered an essential improvement to the developing tourist industry.

A narrow-gauge railway was constructed to run from Aberystwyth to Devil's Bridge to transport ore from the mines of Cwmystwyth, Frongoch, Cwmrheidol and other remaining mining enterprises. The promenade was lit at night with electric light, and there was significant progress in improving the cleanliness of streets in the town with the introduction of a better sewage system and clean water supply. A cliff railway was built at the north end of the promenade to provide access to Constitution Hill. It opened in 1896 and remained the longest funicular railway in the British Isles until the opening of the Cairngorm funicular in Scotland in 1987. The country was changing and Edward VII, the new king, supported the modernisation of society through advancements in technology, as well as a closer relationship with Europe.

With the help of his son Wil, Captain James continued with his interest in gardening and they set aside a part of one of the small fields to grow crops for both their own consumption and also to be sold at the market in town. His meticulous approach to his labours meant he read and researched to make sure that his new venture would be a success, and he wrote letters to enquire about books and catalogues which could help him:

Messers. Sutton & Sons
High Holborn
London
Dear Sirs,
    Being of want of a good book on gardening, I have been recommended to get one from you. Please let me know the price of the one called The Gardener, *then I shall send the money.*
    Yours Truly
    Wm. James
    P.S.
    I will be extremely obliged for one of your catalogues on seed.

As ever, his scrupulous attention to detail and record-keeping meant that this venture was also recorded in his notebooks.

Memorandum of Gardening

5 March	*Sowing a pinch of celery.*
6 March	*Sown a pinch of Kings Cauliflower, also pinch of celery under glass.*
7 March	*Sow 4 rows Ashleaf Potatoes in upper part of field.*
9 March	*Transplanted a few winter onions close to the barn.*
12 March	*Sow Ashleaf potatoes by the little house.* *Weather very wet.*
14 March	*Sow Ashleaf potatoes.*
20 March	*Sow three beds of brown globe onions.*
20 March	*Sow small pinch of all heart cabbage seed and nonpareil cabbage seed.*
21 March	*Sow 3 beds of leeks in upper garden and small pinch lettuce.*

His son William Richard shared his new enthusiasm for gardening, and worked to develop the enterprise and establish markets for their produce in the prospering town. This early interest in gardening would eventually provide him with secure employment in the town's parks and gardens department. The town had become a popular seaside resort for industrial workers from south Wales, the Midlands and the Black Country. This resulted in many dozens of new boarding houses and hotels being set up to meet this increasing demand.

At the age of nineteen Wil had enlisted in the Army and was based in Bedford. He was unable to make it home to support his parents when, in September 1915, they were summoned to appear at the town's Coroner's Court due the unfortunate death of an eleven-year-old boy from Spring Gardens who had died from lockjaw.

The boy, and others, had been playing in the hayloft at Old Ropewalk Farm, and Captain William James, in his statement to the Coroner, said that he was in the kitchen on the morning of 18 August when he heard pitiful crying and saw a boy coming down the steps of the barn. He'd then told the other boys who were with the lad to take him home. The boy's father, a sawyer from Trefechan, took him to the infirmary where he was kept in for the night, but was discharged and continued to be an out-patient until the 31 August. Sadly, the boy died from the effects of lockjaw.

Jane James had stated to the Coroner that she saw nothing during the event as she was working in the hayfield, and denied that she had asked the boys to go into the loft to press down the fresh hay crop. She stated that she knew that the boys were in the habit of playing in the barn. The injury had been caused by an old chaff-cutter which had not been used for years, and they did not know that it still had a handle and a knife.

The Coroner stated that, in the light of the presence of a potentially dangerous implement, the barn should be secured in future.

The matter was made worse for Captain William James and his wife, as a full account of the misfortune appeared in the *Cambrian News*, the local newspaper, on 17 September 1915. But, much to their relief, the concerns within their community over the escalating war deflected some of the inevitable gossip.

2

War

William Richard James
(1895–1968)

WILLIAM AND JANE'S son Wil worked hard on the farm. Aberystwyth had become a lively town with a growing number of tourists. In the summer months large crowds would arrive to walk the promenade and visit the beach and take part in other attractions such as sea bathing, boat trips, fishing, the cliff railway and the Vale of Rheidol Light Railway, as the Devil's Bridge line was known. The Great Western Railway invested in charabancs which operated in the busy months, with trips along the promenade and the nearby coast as well as a six-wheeled vehicle which went to the summit of Pumlumon. On some days thousands of people would arrive on excursion trains from the Midlands and beyond. Occasionally the crowds were so great that notices would be put on the promenade requesting people to keep to the right when walking. The *Cambrian News* even reported that 'The streets were uncomfortably crowded and hundreds of visitors could not get seats and it is getting worse every year'.

However, the national mood was less positive as activity in

Germany was causing increasing anxiety and concern. There were continuous reports in the press during June and July 1914 of increasingly belligerent activity by Germany, who declared war on Russia on 1 August. By 3 August Germany had declared war on France. Germany's invasion of Belgium gave Prime Minister H.H. Asquith no option but to confront this belligerence. In his statement to parliament he said that:

Owing to the summary rejection by the German Government of the request made by His Majesty's Government for assurance that the neutrality of Belgium would be respected, His Majesty's Ambassador in Berlin has received his passport, and his Majesty's Government has declared to the German Government that a state of war exists between Great Britain and Germany as from 11.00p.m. on the 4th. August.

Lord Kitchener was appointed as the Secretary of State for War on 5 August and set about appealing for 10,000 volunteers.

William Richard, at the age of nineteen, had decided to join the Cardiganshire Battery which was known simply as the Battery. It was officially the Cardiganshire Battery, 2nd Welsh Brigade Royal Field Artillery, which was part of the large 53rd Division. The Battery was a volunteer unit which had been formed in Aberystwyth on 28 June 1901 as part of a national reaction to the Boer War, and each Battery was supplied with four 15-pounder guns.

When Wil and his friends joined up, they would have felt that they knew how to handle and carry a gun as they hunted foxes and rabbits on their farms. They, like many others, thought that the war, when it came, would be a short and contained affair and that, by joining up, they would have some respite from the daily routine of farm work or work in the mining industry.

The initial time after accepting their fate within the Battery

The young soldier, the author's grandfather, William Richard James

The camp at Trawsfynydd

William Richard James at the camp in Trawsfynydd

was daunting, as they would be subjected to rigorous training and considerable pressure from the noisy, threatening and swearing NCOs. They were stationed at a training camp in Trawsfynydd in north Wales, established in 1906 by the War Office. The camp owned over 8,000 acres of land which was used for artillery practice by both regular and territorial army. Typically, soldiers stayed for two weeks at the camp during the summer months and would be housed in tents which were laid out in rows, with officers living a respectable distance away from the enlisted men. Here, as well as their initial military training, they would also participate in live fire artillery training. They also learned horsemanship skills and rode and drove a large number of horses which were at the camp. A small permanent village was established to the west of the camp at Bron Aber, where there were shops. The camp was referred to by the men as Tin Town, due to the construction of many buildings using corrugated iron sheets. Generally, at that time of year, life would have been fairly comfortable in the camp.

On completion of their training, they duly returned to Aberystwyth. In the days prior to the declaration of war, the Battery was told to be ready for deployment, and on the day that war was declared they were told to report to the Drill Hall in Glyndwr Road, Aberystwyth. They would all stay at the Alexandra Road School. The Battery then left for Pembroke Dock and were housed in dormitories, but were poorly provisioned and fed. In a postcard home William Richard wrote:

Pembroke Dock, 16 August 1914
Dear Mother
Just a line hoping to find you in good health, as it leaves me at
present. We do not yet have enough food. We have to buy it and we
have no beds. Only one blanket to sleep on. We had some rum on
Saturday.
　　Your loving son
　　William

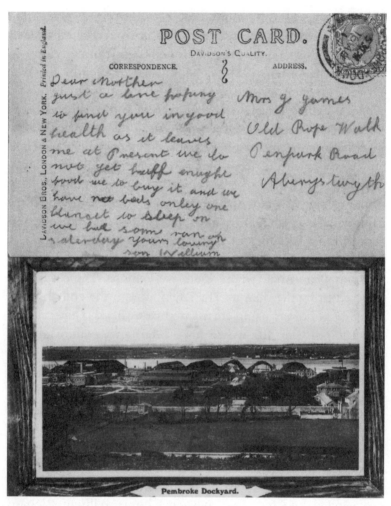

POST CARD.

DAVIDSON'S QUALITY.

CORRESPONDENCE.   ADDRESS.

Dear Mother just a line hoping to find you in good health as it leaves me at Present we do not get half enough food we to buy it and we have no beds only one blanket to sleep on we had some rain on Saterday yours loving son William

Mrs G James Old Rope Walk Penparke Road Aberystwyth

Pembroke Dockyard.

William Richard James's postcard home

He wrote a separate card to his father:

Pembroke Dock, 16 August 1914

Dear Father

Just a line hoping to find you in good health. Did you receive postcard? On top of this hill is the barracks. I have a grand view of

Pembroke Dock. We went to Pembroke town on Friday, 6 miles from
Pembroke Dock. On the guns.
    Your loving son
    William

There was communication between the Battery and Aberystwyth, and by the following day he wrote another postcard home:

Pembroke Dock, 17 September 1914
Dear Parents
Just a line to let you know that I received your letter and very glad
you sent some cakes and tart. Well, you won't know me when I come
home. They say I am getting fat. What do you think of the British
getting 25,000 Germans and five miles of guns? Don't be down-
hearted, I am all right. I am in the blue at the same old job. What do
you think of the cards I sent you?
    Your loving son
    William

The Battery was then told they were being transferred to Northampton to continue training, and Wil wrote a postcard home informing his parents:

Pembroke Dock, 17 October 1914
Dear Parents
Just a few lines to let you know that we are going to Northampton
sometime next week. I don't know if I am coming home on Sunday.
    William

On 18 November Wil and the rest of the Battery were warned that they were to be posted to India, but this order was cancelled and they were moved to Cambridge, where Christmas was spent. But in 1915 they were on the move again, going to Bedford.

Northampton, 11 August 1915
Dear Parents
Just a line to let you know that I arrived in Northampton and I stopped there for six hours. I leave Northampton at nine tonight for Bedford. I do hope you don't worry about me. I will be home, free leave, in a month or two.
William

Wil was having some difficulty with his pay, which he was unable to sort out. He wrote to his father in the hope that his well-established connections back in Aberystwyth could help resolve the issues.

Bedford, 11 August 1915
Dear Father
Just a few lines to let you know that I arrived safely. Have you sorted with the pay yet? Try, try again with the big ones in Aber. Like Mathias, you know who I mean. I only have one parcel with me going back. I hope this will find you in best of health.
Wil

Major Mathias was the recruiting officer for Cardiganshire, and whose family owned the Cardiganshire Steam Navigation Company. He would have been well known to his father through his maritime activity.

The Battery was part of the 53rd Welsh Divisional Artillery ordered to France to join the British Expeditionary Force on the Western Front in late November 1915. They were initially sent to various locations for instruction on front-line duties. However, fortunately for Wil, they were not deployed to the Front but were ordered south, to Marseilles, from where they would cross the Mediterranean and sail on to Alexandria in Egypt, where they arrived on 11 February 1916.

They then moved to a camp north-west of Cairo, on the edge of the desert. Here they experienced sandstorms, and had to remain in their tents which made feeding and watering the horses difficult. Life in Egypt was different and new to country boys from Cardiganshire. They had to come to terms with sand-flies, sandstorms, camels, and especially the heat, and weeks and months without rain. They had to adjust to wearing their heavy cotton khaki uniforms, pith helmets and the laborious process of putting on puttees. Food was not in short supply, but they moaned about the monotony of a diet of rice and beef and a limited supply of water. St David's Day 1916 was celebrated in a different way, with camel and donkey races. Their greatest source of comfort was their cigarettes, or 'rattlers' as they were known from an old army saying – they could have a 'rattling good smoke'. Cigarettes and a good smoke

William Richard James (right) with two companions at the Pyramids in Giza, 1916

also had the advantage of keeping the flies away. They had opportunities to visit and see the Pyramids and the Sphinx, things which they had only read of or were told about while at school.

In April 1916 they were part of the 53rd Division guarding a section of the Suez Canal. They had the advantage of being able to bathe at least once if not twice a day in the heat. On 20 January 1917 the whole Division commenced a long, hot and dusty march to el-'Arīsh on the Mediterranean coast. The Egyptian Expeditionary Force had launched the Sinai and Palestine Campaign by crossing the Sinai Desert and advancing against the Turkish forces.

They became involved in the second battle of Gaza, and by late August were resting after five months of activity without a break in a summer which had been long and hot with temperatures of up to 48 degrees Celsius. The third battle of Gaza followed. By 31 October they had moved on and captured the wells of Beersheba in the Negev Desert, having marched for days on roads which were no more than dusty tracks. They moved into the hills north of Beersheba, together with the Camel Corps.

Crossing the Sinai Desert into Palestine demanded considerable support, especially the need for water in an area where there was little if any available. Water was piped from the Suez Canal to near Beersheba, a distance of over 147 miles, and engineers also laid tracks so that water trains could supply the Front. These water trains, in turn, supplied camel trains which carried the water in bags to supply the troops.

The attack on Beersheba had taken the Turks by surprise, and had succeeded with only slight losses to the British force. Around 2,000 Turkish prisoners were taken and the capture of thirteen guns. However, around 500 Turkish soldiers were buried in mass graves in the desert. This success improved the access to a good water supply. There were seventeen wells in Beersheba, many of which had been known since the days

This photograph section contains sixteen paintings
from the hand of Captain William James.

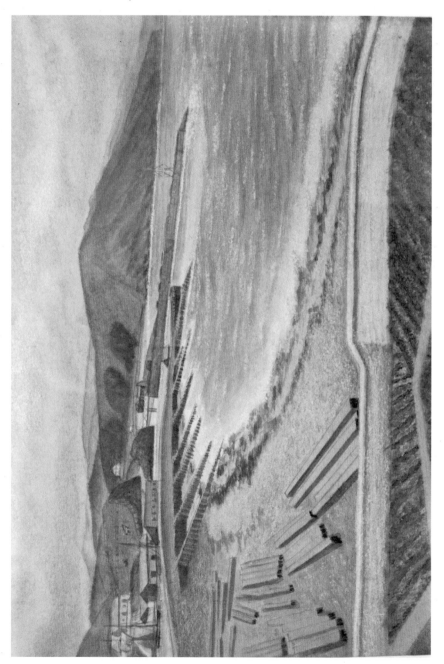

Imported timber on Ro Fawr beach.

View across Ro Fawr before the promenade extension of 1901.

View across Ro Fawr and the harbour.

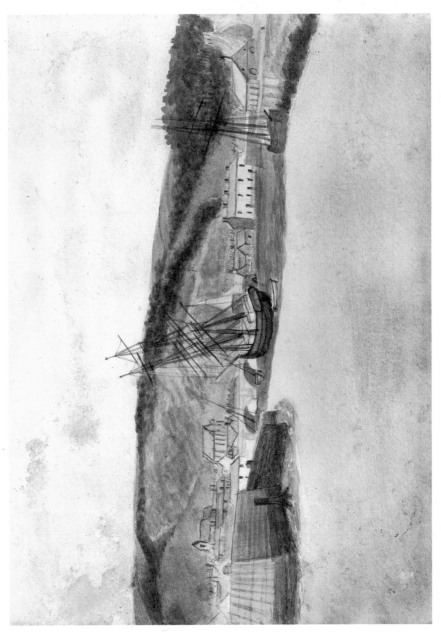

Nineteenth-century Aberystwyth harbour, looking upstream to Trefechan bridge.

Schooner SS *Helena* entering the harbour.

Old Ropewalk farm buildings in the foreground. The main road can be seen below the line of trees and the castle in the far distant left.

Ship moored in the harbour.

Ship sailing north from Aberystwyth, late 19th century. The pier can be identified; this was built in 1864.

Vessel approaching the harbour. The ship is depicted perilously close to the stone pier. The black ball on the mast on the pier was hoisted to indicate that there was nine feet of water over the sand bar. Hobblers are waiting on the pier to offer assistance into the harbour.

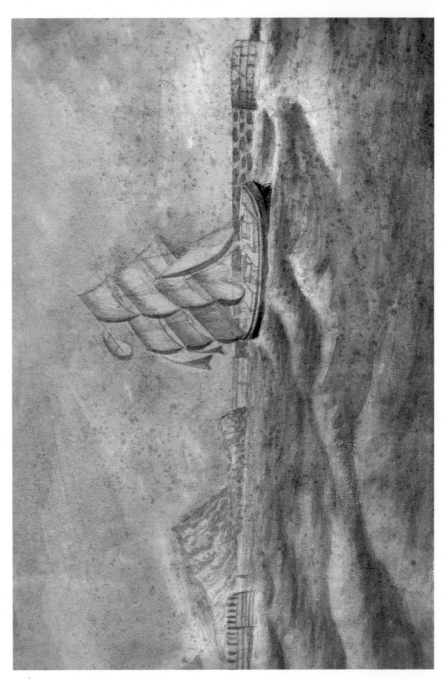

Ship in full sail heading into the harbour.

Harbour, late 19th century. The two pillars on the hill were used as sighting posts for ships approaching the harbour entrance.

Maritime activity below Constitution Hill, late 19th century, showing a steam powered craft and a lifeboat.

Part of Old Ropewalk farm. Painting signed and dated. This is the same gateway depicted in the photograph on page 52.

Sailing ship in full sail.

Small sailing vessels offshore. No south promenade and Pen Dinas in background.

A classic view of Aberystwyth from Constitution Hill. The length of the pier is noteworthy and the two sighting posts for ships to use to enter the harbour can be clearly seen.

of Abraham in the Old Testament. Further north there were remote wells of 100-foot and 210-foot depth, and Gaza had an abundant supply of wells. Large convoys of camels would transport water to the front line.

Despite his isolation in the Sinai Peninsula, Wil would still have some communication from home. The Army Postal Service of the Egyptian Expeditionary Force had an efficient network which was able to deliver up to 2,500 letters a week, together with over 2,000 parcels. These parcels were a significant morale booster to the troops, and many were arranged by the Comfort Fund back home. They would often contain home-knitted socks and other welcome goods, but the most appreciated content was the cigarettes.

On 5 November 1917 they were shelled constantly throughout the day, but held their ground and the enemy was finally driven away by accurate artillery fire. Further severe fighting took place the next day, and they improved their position by taking control of several hills as well as taking several hundred Turkish prisoners and securing a number of their guns. Progress was slow however, and the troops suffered considerably from thirst, as an exhausting hot, dry wind blew for two days.

The Welsh boys were engrossed in the day-to-day activity of battle and survival, but were also fascinated and confused by the fact that they were operating in an area of biblical significance, as the place-names of the towns and villages through which they passed were familiar to them through their regular attendance of chapel and Sunday school back home.

In fifteen days the Egyptian Expeditionary Force had advanced sixty miles and removed the Turks from an area which they had occupied for more than six months. On 4 December Wil's Division commenced a long march towards Jerusalem. Hebron was taken on 5 December and Bethlehem was occupied by 9 December. They soon captured the Mount of Olives, where they stopped for a well-deserved rest, but influenza was causing many non-combative casualties. Jerusalem was captured on 9 December, and on 11 December General Sir Edmund Allenby, the Commander-in-Chief of the Egyptian Expeditionary Force, made his official entry into the city on foot through the Jaffa Gate. He highly praised his men and stated that they 'had acquitted themselves in a manner beyond praise'.

The large force now amassing in Jerusalem required 400,000 gallons of water a day, and some of the water supply systems in the city dated back to the Roman days of Pontius Pilate and Herod. Water travelled through rock-cut channels and masonry aqueducts, and the engineers of the Egypt Expeditionary Force cleaned these out in order to fill reservoirs from which they could run water pipes for both military and civilian use. Not since Roman days had there been such a plentiful supply of water in the Holy City.

On 27 December the Turks embarked upon a sustained attack on Jerusalem, but were repulsed with heavy losses and retreated.

Wil's Division advanced north in heavy rain, and experienced continuous action throughout February 1918, but their position in the area was stabilised and they remained in the sector throughout the summer.

By 21 September the 53rd Division was still fighting the enemy, even though the Turks were retreating. However, by the early hours of 4 October the German Chancellor, Max Von Baden, who had been appointed by Kaiser Wilhelm II, had sent a telegraph message to President Woodrow Wilson in the United States requesting an Armistice between Germany and the Allied powers.

Armistice was declared on 31 October 1918, and finally hostilities came to an end on 11 November and it was time to think of home.

The scale and size of the operation they had been involved in was enormous. They had used 32,712 camels, as well as thousands of donkeys, the sky-blue galabieh of the camel drivers contrasting starkly with the drab khaki of their uniforms. This local form of transport had been supplemented by 1,579 lorries, 1,523 motorcycles and 586 ambulances. Crucially, they had also used 281 caterpillar tractors which had been vital in ensuring movement during the winter months. In November 1918 they moved to Alexandria, but did not leave until March 1919 for their long journey home to Aberystwyth. They had fought and advanced uncomplainingly in a hostile and unfamiliar land, experiencing thirst, heat, cold and fatigue.

Demobilisation officially began on 20 December 1918 and was completed by June 1919. By July 1919 they were all home and the Battery was placed in suspended animation, to be reformed in 1920. They would be changed men for ever. The innocent, naïve country boys who left home in 1914 would return older, and with a different view on life and their position in it. They were led to believe that, as a result of what had happened, they stood on the threshold of a new age with a fairer and more equal society to look forward to. Many families had to come to terms with the loss of fathers, husbands, sons and fiancés. The majority of those who had died were buried in foreign fields, which deepened the grief and obstructed closure.

Wil's father, Captain William James, had died on 19 August 1917, aged seventy-five. Wil's placement at the time in the Sinai Peninsula meant he had been unable to get home, but was informed of all that had happened via letter. His father's obituary was published in the *Cambrian News*.

DEATH
In the death of Captain William James, a link is broken with the old shipbuilding days of Aberystwyth. The deceased was a son of Captain Wm. James of the schooner Margaret Evans, *the first Aberystwyth ship built by John Evans senior. Capt. James born on March 15th 1842, was educated at Skinner Street school by the late Rev. Wm. Jones and Mr. John Evans. He commanded several vessels – the* Margaret Evans, Helena, William and Mary, *and* Eigen, *the*

latter being lost at Casablanca. He was twice married, his first wife being Miss Mary Evans, daughter of the late Mr. Wm. Evans, Rope-house Fach. Subsequently he assisted his wife in the milk business. His second wife was Miss Jane Williams, daughter of the late Mr. Rd. Williams, Tanllan Mill. He had one son, who is at present in Egypt. Captain James was an artist and painted many pictures of local interest. He was a man of sterling character and a member of Tabernacle Church.

Wil's mother had sold the farm as a going concern, as she was a slight and frail woman who could not manage the farm on her own and did not know for certain when, or if, her son would be home. War had caused a change in agricultural practice, with a focus on corn production. The amount of land used for grazing and hay fell, but there was a recognition in the value of farm work, with increases in wages. Jane James purchased a small terraced house in Edge Hill Road overlooking Aberystwyth, and it was to here that Wil returned in 1919.

He had to seek employment, and his interest in gardening encouraged him to get work as a labourer and gardener in the town's Corporation workforce. This enabled him to be out and about in the town, which had changed substantially since he left in 1914. The previous strict chapel-controlled life seemed a little more relaxed, and the number of visitors in the spring and summer months meant that the town was busy and bustling.

Wales had changed too through the experience of war. Young men who had previously lived quiet secluded lives had been exposed to a wide world and had experienced unspeakable horrors. Many had fought in proudly Welsh divisions within the British army, which had enhanced their national identity and would, in due course, establish and advance the cause of Welsh nationalism.

The war had been a watershed moment between the virtual demise of rural neighbourhood living and an industrial society in which the products of a modern urban culture had become known to the many and not the few. This filtered down to seaside towns like Aberystwyth through the continuing development of rail links.

Wil met a tall and elegant girl whose family lived in a three-storey house in Bridge Street, Aberystwyth. Lizzie Jenkins lived with her parents, and her father was employed as a chauffer. Richard Jenkins was a tall, striking man, and his wife Helena Jenkins, whose family came from Llanfihangel-y-Creuddyn, was also a tall, imposing woman.

Wil and his girlfriend spent a lot of time together in Aberystwyth which was busy, vibrant and a popular resort with a pier, bandstands, ice rink and several places of entertainment. They would walk the promenade and go out to the remoter beach at Tanybwlch or wander up Pen Dinas hill where the memorial to the Battle of Waterloo had been installed – an imposing upturned cannon.

Their relationship blossomed and Wil's newfound independence and confidence, together with the warmth of a fine summer and the solitude they often sought, resulted in the consummation of their feelings for each other. Lizzie became pregnant and they both had to go and face her family. Life would change for both of them. They were married in the town's Registry Office on 17 December 1920. Wil was now twenty-five, and Lizzie younger at twenty-two. It was a quiet ceremony, and a pregnancy out of wedlock was an embarrassment to both families who were respected members of Tabernacle chapel. They knew that tongues would wag in the town. They moved in with Lizzie's parents, and on 2 March 1921 their daughter Helena was born, and was named after Lizzie's mother. Wil's mother would now live alone in Edge Hill Road. On 22 January 1923 their second daughter, Jane, was born and life in Bridge Street became crowded with a family of four, as well as the older parents.

Wil and his young family moved back to live with his mother in her terraced house, and he worked hard to ensure that his new family was well provided and cared for. The town was a pleasant place to bring up children, especially in the summer months when it would be full of visitors. Wil and Lizzie always made sure that they were well turned out, and when the family would be out in town they would be well dressed. They would often have days out, visiting areas outside Aberystwyth. Charabanc tours were available from the Jones Bros. and West Wales Garages of North Parade, and Primrose Motor Co. of Terrace Road which, for a modest charge, would take visitors to Llandrindod Wells, the Elan Valley, Talyllyn lake and beyond. Aberystwyth was promoted as the Biarritz of Wales, and those coming to shop in its variety of businesses were advised to post their orders prior to their visit to avoid delays.

However, there was discontent in the countryside and farmers were unhappy. During the war they had been asked to increase production, especially of corn, and had been given an undertaking by the Government in 1917 that the price would be guaranteed for six years and that there would be minimum rates of pay for farm workers. The Government had repealed on this promise, and the value of farmland had started to decline. By 1921 an agricultural labourer earned 44s. 2d. per week, but this declined to 28s. per week by 1924.

In 1924 a war memorial to the fallen of the town and surrounding area was officially unveiled in an emotional ceremony. The imposing bronze statue at Castle Point arrived by sea from the sculptor's studio in Italy. It was erected by a workforce of unemployed married ex-servicemen. The economic crash in America eliminated demand for British-made goods, so King George V urged Prime Minister Ramsay MacDonald to form a National Government. The election of 1931 resulted in a Conservative-led Government, but despite the slump and unemployment of the Depression, Wil managed to stay employed and his family prospered.

Their daughters grew into attractive teenagers, and would be entrusted to go into town and walk the promenade in each

William Richard and Lizzie James with their daughters. Jane James and Helena (Lena), the author's mother, on the right.

other's company. It was here that they met boys. Eldest daughter Lena, as she was known, worked as a shop assistant in Mrs Dewar's ladies wear shop in North Parade. She frequently met with a young man who worked at a shop in Great Darkgate Street.

Elfed Davies was from Salem, a small hamlet outside Aberystwyth, and he was one of seven children, but the only boy. Due to the distance of Salem from the town, and the six-day week, he lodged in town with a relative. Their time together was carefree, and at weekends they would catch a bus on Saturday afternoon and travel up to Salem to stay with his parents and go for walks beside the small lakes in the hills which lay above Salem.

Lena's sister Jane had also taken up with a boy whom she had met in town, but she was a livelier girl and, to her sister's dismay, she would frequently have two boyfriends at the same time. Jane was courted by Owen Jenkins, a young man from the secluded hamlet of Aberffrwd in the Rheidol valley. He worked as a footplate fireman on the railway which ran between Aberystwyth and Devil's Bridge (Pontarfynach in Welsh) and would be able to travel for free between the town and his isolated home.

The girls and their respective partners enjoyed the lively

town, especially in the summer. Large numbers of boats would be on the waterfront close to the imposing building of the King's Hall, which had opened in 1933 as an Art Deco dance hall and provided music, dances, amusements and indoor dodgems in the basement.

During this carefree time, however, there was disquiet on the Continent. The Chancellor in Germany, Adolf Hitler, had invaded Austria and Czechoslovakia and was intent on expanding his Nazi empire. Elfed was required to register for National Service, and by September 1939 the carefree days were over and he left for a military training camp. In July 1941 Lena volunteered for service in the Women's Auxiliary Air Force (WAAF) and went to Blackpool for training. She had to adjust from living in a cosy terraced house in Aberystwyth to communal living in a Nissen hut at an RAF camp. From there she went on to various RAF stations in Oxfordshire and spent a considerable time at RAF Bicester. This station was a base for the Bristol Blenheim bombers and also the Handley Page Halifax bombers. It became the basic training base for British and Commonwealth aircrew, and Lena would meet a lot of airmen during their short stays at the base. She was then posted to RAF Abingdon which was a Bomber Command station at which Wellingtons and Lancasters were based. She was involved in the repair of aircraft which had been on active service, and would often be required to fly in them on completion of the service work.

Elfed's father died in March 1942 and he returned home for the funeral at Salem chapel. John David Davies was interred with his daughter Hannah Mary. Elfed was aware that he was to be posted abroad on active service to join the North African campaign. He and Lena decided that in the circumstances, as they had already got engaged, they would get married quickly and quietly, respecting his mother's recent loss. In true Cardiganshire tradition they could take advantage of the married allowance while away to accumulate a nest egg for their new married life on his return. They were married in a simple ceremony at Tabernacle chapel on 14 April 1942, with the Rev. J.E. Meredith officiating.

Lena's sister Jane signed up for the Women's Land Army, and would work in the Aberystwyth area, thus being able to remain in contact with Owen who was still employed in essential non-military work on the railway. Work in the Land Army was hard, and Jane came into contact with a wide range of girls who had been posted into the area and were living in a hostel outside the town. This widened her life experience after her relatively sheltered upbringing in the small seaside town. They were paid 22s. 6d. a week in wages, but were often taken advantage of by some farmers who gave them dirty and difficult jobs to do. In due course farmers realised that the girls were fit and capable of doing men's work, and they could operate threshing machines and other farm implements. Jane enjoyed her work and the security of still being in her home area and maintaining her contact with Owen, in contrast to the lengthy periods of separation experienced by her sister.

Lena (my mother, second left) in WAAF uniform, Jane in Land Army uniform and their father, Wil, in a Civil Defence uniform

Wil was too old for active service but he enlisted in the Civil Defence Volunteers which became the Home Guard. He served in the 1st Cardiganshire Home Guard and would join his fellow volunteers at Pen-yr-angor where a rifle and clay pigeon shooting range had been established for practice. The Home Guard remained in operation until February 1947.

Jane married Owen at Tabernacle chapel on 25 January 1946. I was born on 13 June 1946. The wartime years had passed with no major incidents for the family and they, as changed people, would face the challenge of a different life in a post-war Wales.

3

Mine

John David Davies (1876–1942)

My GREAT-GRANDFATHER ON my paternal side lived in the Lledrod area in the hills between Aberystwyth and Devil's Bridge. He was a lead miner who earned his living in the mines at Frongoch, near Trisant, but he had been unwilling for his son John David Davies to follow him underground. Mining was a laborious and dangerous occupation, with frequent injuries, and health was damaged by the cold, damp conditions and the effects of explosive vapours resulting from poor ventilation.

He arranged for his son to work as a farmhand, or *gwas fferm*, at Llain Fawr farm which was a simple enterprise nestling close to Devil's Bridge. John Davies's wages would be meagre, earning but a few shillings as the bulk of the reward for his labour was simple accommodation in a loft over the barn and being fed thrice daily alongside the owner and his family. Employment of farmhands was a common way of providing work as well as enabling large, poorer families to relieve the burden on both accommodation and expense.

Life as a farmhand was tough and physically demanding.

The seasonal work pattern varied, from peat cutting to the hay harvest. Peat cutting took place in the hills beyond the farm and the long narrow turves would be cut and allowed to dry in the sun and wind before being gathered and carted down on a simple drag cart and stacked close to the house to provide fuel for the winter months. Other days would be spent in the surrounding fields cutting hay and corn by scythe, which was also physically demanding work. The cut hay or corn would also be stacked into bundles or stoops, and placed at regular linear intervals along the fields and allowed to dry in the sun before being transported back for winter storage or threshing.

After a few years John Davies became restless, and persuaded his parents that he could attend the November fair at Aberystwyth where there was still a tradition of employers meeting potential servants to secure employment for the coming year. The agricultural hiring fairs had become less popular than they once were, as many had gone to work in the lead mining industry, and others had gone south for more lucrative employment in the coalfields of south Wales.

John Davies was successful in striking a bargain with John Jones of Tynpynfarch, a farm on the outskirts of the village of Penrhyncoch across the valley from the substantial farm of Brogynin Fawr. He would start in the new year. He moved to Tynpynfarch and settled in with the large family who lived there. Here he met and courted a young maid servant who also lived and worked on the farm. Marged Ann Morgan was a tall, slim, statuesque girl whose family lived on the hill two valleys away in the hamlet of Cefn Llwyd, close to the productive lead mines at Daren and Cwmerfyn. Her parents had negotiated a period of service for her, as was often the case for young girls in the area. It meant one less mouth to feed.

John Davies and Marged Ann would spend what little spare time they had walking fields beside a small stream. They attended chapel together and would occasionally visit Aberystwyth, as well as attending local concerts or gatherings

The young John David Davies. Photograph taken by L. Morgan, Borth.

which would take place in Penrhyncoch or high on the hill at Bontgoch. John eventually proposed and she accepted. Although Marged Ann was nine years younger than him, John Davies was careful not to offend her family and sought permission for her hand in marriage from Marged Ann's father, William Morgan of Cefn Llwyd. He had been supported by his employers, John and Catherine Jones of Tynpynfarch, and had discussed his intentions with his parents. They persuaded him to return to Llain Fawr for a period of six months before the wedding in an effort to save as much money as he could.

John Davies and Marged Ann were married by the Rev. David Lewis at the Methodist chapel in Capel Madog near the village of Cefn Llwyd on 21 December 1906. Both families attended. His father, Edward Davies, was still a lead miner working at Frongoch. Her father, William Morgan, worked locally as a farm labourer and was pleased that one of his daughters would now be settled and provided for.

John Davies eventually decided that he would need to find a job with more income in order to support his new circumstances. So he decided to seek employment in the mines despite the earlier objection of his father. His standing in the locality as an honest and reliable worker enabled him to secure the rental of a small end-of-terrace cottage in the hamlet of Salem which stood on the hill overlooking the valley. The house was a simple stone-built property, with a slate roof and a sizeable garden to one side.

Salem was a small, linear hamlet, its approach guarded by the large imposing mass of the Manse. The Manse was set back respectfully from the road and was protected by a hedge growing out of a high stone wall. The approach was through a small iron gate and up a narrow path, and there was no way of approaching unseen towards the front door, as would have befitted the most respected member of the community.

Progressing past the gate, and on the right beside the road was a small open compound surrounded on three sides by a crumbling moss covered wall. The site was overlooked by a galvanised, corrugated-sheet shed which overlooked the steep fields descending to the floor of the valley and the Stewi stream. The compound was the communal property of the first three houses.

John Davies's garden had a well-kept path, which led directly up the gentle slope to a tall, small wooden closet, complete with a square timber seat below which stood a substantial galvanised steel bucket. This was the only toilet available to the household. Alongside the gate was a very large steel drum into which was channelled water from the downpipe draining the neatly painted gutters below the slate-topped eaves of the cottage. There was no running water in the house, and water from the garden butt would be used on washing day. Rainwater ensured softness, thereby economising on the quantity of soap required. Drinking water was carried from a pump located a further hundred yards beyond the terrace.

The single living room was dark; cool in summer and warm in winter. The focal-point fireplace was large and open, with a strong swinging arm and chain on which hung a large black cast kettle. The black-leaded hearth had an assortment of strong, long-handled implements, and there was an adjacent wall oven which could be used on bread baking days.

The cosy lived-in room led to the scullery, where large hooks protruded down from the ceiling from which suspended hams and bacon from the annual slaughter of the communal pig. A steep stairway led to an open landing bedroom which, in turn, led via a firm panelled door to the one large bedroom.

Beside the house was a small shed which contained the coal and wood for the fire.

John Davies worked at Camdwr mine, which was situated a considerable distance from Salem and his new home. However, the guarantee of a job and regular income had to be taken, and he was prepared for the long walk that would be necessary. Despite being thirty years old, his lack of mining experience would mean that he did not go underground and he had to be satisfied initially with work above ground at what was known by the miners as the flooring, or *ffloriwn* in Welsh. This was where the extracted ore was crushed and sorted to remove the useless debris and isolate the valuable mineral ores.

The material which came out of the mine was transported in drams, which delivered the mixture of rock and dust to the stone breaker. Here it was crushed by thirty-inch cast-iron rollers which were powered by a forty-foot waterwheel. It was then sorted by a jigger, which was a large, rectangular multilayered sieve which sorted the crushed rock. It was then washed by water during the process. The water used carried much fine ore, and this sludge was allowed to flow through a number of slime pits of rectangular tanks. The sludge would then also be reworked to ensure that all the ore was recovered. It was hard and dirty work, which wetted and chapped hands, and chilled the body. The short lunch break provided welcome relief. The men who worked underground were paid more and were also out of the bad weather in winter. But they also suffered from the cold and wet, despite the fact that their daily hours were two hours shorter than those who toiled on the surface. John's days were hard and the long walks to work and back extended the time he was away from home. His strength and hard work again endeared him to his work colleagues, and they were eager to educate him in the skills of mining as well as trying to explain how to be at one with the rock and learn how to read the signs and signals of a favourable seam and be prepared to understand the mystery of the 'knockers'.

The old miners firmly believed that there were beings from the world of fairies occasionally prepared to help them whilst working underground. They spoke of the tapping noises similar to a chisel or pick being struck, and that the noise would lead them to a new vein of ore. Mining in the area generally had lodes which dipped at between seventy to eighty degrees, and ore was removed via adits rather than shafts.

His wages were paid monthly, and by the end of February 1907 he was able to purchase furniture for the house. They took delivery of an oak dresser, a table and a small table made

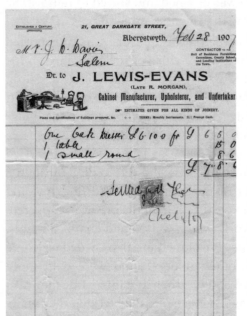

by J. Lewis-Evans, a cabinet manufacturer and undertaker of Great Darkgate Street, Aberystwyth, and by the end of April he had paid the debt off in full.

The mine, however, was struggling to keep going and John Davies was forced to seek new employment. He had been told that a mine in another valley, above Talybont, was being reworked and he walked over to see if could find work. His initiative paid off and he was employed as a miner and would be going underground. By comparison this was a much more modern mine.

Bwlchglas had been mined since the early part of the 19th century and, like many other ventures in the area, had been subjected to changes in its fortunes. The remote valley where the mine was located had a varied history. In addition to lead mining there was quarrying activity at the isolated Cerrig yr Hafan. The quarry was located in a high position at the head of the valley, and was worked for the extraction of a coarse textured sandstone. It yielded 1,690 tons in the form, mostly, of paving sets between 1897 and 1898. Some of the stone was used in the construction of the sea wall and front in Aberystwyth.

In an attempt to facilitate both the transport of rock and ore from the mine, a narrow-gauge railway was constructed in 1896 and completed a year later. It became known as the Plynlimon and Hafan Tramway and ran between the village of Llandre (Llanfihangel Genau'r Glyn), past Talybont, and on up to the head of the valley where a steep tramway linked it to the quarry. However, it was not a success and closed in 1899, despite efforts to run it as a tourist attraction. Some of the engines and rolling stock were eventually sold to the Vale of Rheidol Light Railway which began operation in 1902.

A new group of investors had taken over Bwlchglas mine, and the Scottish Cardigan Lead Mines Ltd. Company put a lot of money into the venture. It was to start in a modest way with a workforce of twelve miners working underground and three above, but the company also employed sub-contactors who were prepared to work long hours. Soon many of the machines were running for twenty-four hours a day.

John Davies's first time underground would have been traumatic but exciting. He would have collected a numbered pick and chisel at the smithy, secured them to his back with the strong cord provided, put his lunch food into his pockets and walked into the level to the shaft. The shaft trap door would be lifted and the 'bargain' team he was now part of descended the wooden ladders held in place by strong iron stakes. He would have followed the miner in front of him, held firmly on to the wooden rungs as both hands were free – his candle was on the rim of the leather helmet on his head. Water dripped down the twelve-foot-square shaft. Once on the level where his bargain gang were working, he'd wait for the others. Some descended the shaft whistling, others singing, but a first descent would be a shivering, fearful experience.

In a small alcove in the rock there would be wooden planks on which to sit, and miners left their lunch packs there before entering the level and workface of the ore seam. The alcove would be a meeting point over lunch breaks and while waiting for the air to clear after the firing of gelignite. Conversation at break times would range from politics, religion and theology to local information and an opportunity to be more informed about mining from the experienced older men.

John Davies worked at the face of the ore seam with pick and chisel. Boreholes would be filled with gelignite, and a safety fuse. When firing took place, the team would hastily retreat to the safety of the rock alcove and wait for the dust to settle before returning to work. He soon learnt that some seams were easy to follow and yielded straightforward income. Other seams proved difficult and would run out, resulting in dangerous work. The attitude was that miners were plentiful and labour cheap, and that there was a 'take it or leave it' approach adopted by the Mine Captain on behalf of his paymasters. Miners were fortunate at Bwlchglas as the Mine Captain was a fair man. When work was scarce, miners depended on the Mine Captain for employment, and often 'favours' were necessary to ensure selection to work. It could be supplying the gift of a rabbit, chicken or some spare ribs from a recently butchered pig.

Some of the hard-working sub-contractors in the mine profited from their long and dangerous hours and earned up to £18 a week, which was almost seventy times the agricultural wage at the time. John Davies's earnings were more modest, but were certainly an improvement on what he was earning as a farm servant. However, the distance of the mine from Salem and the hours he was required to work meant that he would be away from home for five nights a week. He stayed in a barracks which was located further up the valley, some fifty yards from the mine workings.

Very early on a Monday morning he would leave the house with a wicker basket over his shoulder. This contained his food for the week ahead: bread, meat, cheese, butter, tea, and a little bacon, as well as a change of clothes. He would walk wearing his working clothes of moleskin trousers and waistcoat, a wool flannel shirt and a vest and long johns underneath. He wore leather nailed boots, with thick home-knitted socks to keep his feet warm. An old jacket no longer suitable for Sundays was worn over the top, and secured with a scarf or woollen cravat, and topped off with a tweed cap which would be exchanged for a leather mining hat when going underground. In his pocket he would have a fine-knitted balaclava similar to an old-style nightcap, which could add to the comfort of the long walk on a cold day and additional protection when working.

Life at the barracks was primitive and uncomfortable. It was overcrowded and flea-infested, as well as constantly clammy from the damp clothes of the miners returning from their day's labour. They slept six or more to a room. In the winter months, evenings were passed playing cards or having serious political or theological discussions. In late spring and summer their diet could be supplemented by going fishing for wild brown trout in the mountain streams, and in early autumn blackberries would be collected from the hedgerows.

The day was divided into working shifts, with each shift being eight hours. There were two shifts a day and sometimes three, with no time to fully ventilate the mine. Working conditions were poor despite the investment of modern equipment. The mine became equipped with electricity and compressed air machinery, which enabled drilling to take place more effectively. The drill became known as the 'widow maker', because workers returned to the face before the dust had settled and the resulting dust would have a detrimental effect on their health.

Often, the departing shift would have fired a blast and the oncoming shift would enter the workface before the dust had cleared. Bwlchglas was a wet mine, with poor ventilation and low oxygen levels, which added to the health hazards of work underground. As with many other mines in the north of the county, there were also harmful effects to the surrounding land. Water from the mines spread sterility over the fields and meadows and killed fish in the rivers. This was made worse by the extensive system of leats or manmade water courses, which carried water to generate power often up to eight miles away from the source of the water. These leats would overflow in heavy rain periods, and further poison the land.

By 1908 Marged Ann was pregnant with their first child. The fortunes of the mine were still improving, with thirty-two miners working underground. On 10 February 1909 their daughter Sarah, to be known as Sal, was born, and there were new demands on securing a regular wage. The mine thankfully continued to prosper and the labour force had now increased to 304, of whom 135 worked underground.

There was a surge in demand for lead, zinc and silver in the area. Some of the other mines were seeing a revival in their fortunes, to such an extent that the population of some nearby villages, such as Bontgoch, saw a significant rise.

This prosperity had a beneficial effect within communities – the chapels thrived with large congregations. Mine owners and wealthy farmers attended church alongside local politicians and dignitaries, but society was divided and the merchants and smaller farmers attended some of the large Methodist chapels, while the toiling miners and labourers flocked to their own simple local places of worship. Sundays would be strictly observed, with no work done. The vibrant Revival movement, under the influential leadership of Evan Roberts, had a commanding impact on Sunday observance. The large chapel at Salem would be well attended, and the congregation would walk twice on Sunday to worship under the zealous guidance of the Rev. Llewelyn Morgan. Children would attend the Sunday school in the afternoon. As no work would be done on Sundays, water for the family's needs would be drawn from the village pump on Saturday afternoon or evening, and brought into the house to be covered with muslin cloths.

John Davies would always observe the Sabbath strictly. He would recall the predicament of a fellow miner whilst working at Camdwr. This poor man lived in Ystumtuen and, like many others, would stay in the barracks during the week and go home on Saturday afternoon. However, on one occasion, a problem arose at the mine and he had been required to stay on to sort out the difficulties. He was very late setting off for the long walk home in poor weather, and he also suffered from ill health with a chest problem which was common to many working in the dust underground. He eventually arrived home, but after midnight, and when he attended chapel on the Sunday he was approached by the deacons and asked to explain his activity on a Sunday, on threat of being excluded from chapel in the future.

Marged Ann looked after the children and she also had an agreement with a local lady, Mrs Smith, who was also the owner of their house and to whom they paid rent, to look after her chickens and a small flock of geese. The birds were shared, as were the eggs, and on special occasions a chicken would be sacrificed for the table. She also worked hard in the large garden beside the house, and any free time would see her knitting socks for her husband or making simple clothes for her daughter on an old Singer sewing machine, often using the washed cotton cloth of flour sacks.

By April 1911 their second daughter, Lizzie, was born.

Worryingly, the mine operating company seemed to be in financial difficulty, and by 1912 the workforce had been reduced to somewhere near seventy. Working relationships, as well as conditions, deteriorated, with acrimonious reports of minor thefts of nails, and finally culminating in a fatality. Despite this uncertainty, John Davies's family expanded with the birth of another daughter, Beattie (Beatrice) in March 1913. The working week was reduced from five and a half days to four, and the number of employees declined to twenty-one, but he was still able to secure a living wage. The mine temporarily closed in April 1914, which proved a serious hardship for the family, but John Davies succeeded in supporting the home by working as a casual labourer on farms and taking any form of work which was offered, as well as ensuring that the garden yielded as much as possible. The mine reopened in September, which was a blessing for the Salem family as their fourth daughter, Glenys, was born that same month. The following year work was again secure, but by March 1916 most of the men were laid-off, despite a short-lived spike in demand for lead with the outbreak of war.

John Davies's situation became desperate as their fifth child was also born that month. Hannah Mary was a sickly child and seemed unable to thrive like her sisters within the frugal living of the family.

Many other mines in the area were experiencing considerable difficulty, and young men from the villages and towns left to join the war effort. John Davies would be forty in June, but was lucky that he was still in good health. He had to come to a difficult decision. There was no alternative but to leave his home, family and area to go south to the coalfields of south Wales where his skills could be used once more to a profitable advantage. Many miners in the area could be heard saying that it was 'off to Merthyr, Treorci or Nant y Moel in the morning'.

The journey involved travelling to Bow Street or

Aberystwyth to catch a train, taking with them a box of clothes. These working clothes were clean, having been unused for some weeks or even months before the desperate decision to depart had been made. There were many tears on doorsteps, and some of the wives and children would travel to Bow Street and stand on the platform and would continue to wave until only the plume of smoke from the steam train could be seen. A sorrowful walk home then followed to face an uncertain future without a regular income or the support of the head of the family. Often, the eldest sons of some households would accompany their fathers, which would increase the feeling of isolation and worry for the stalwart mothers.

John Davies and his fellow miners would arrive in Bridgend and then travel up the Ogmore Valley where a chain of small mining towns led past Pricetown and on to Nant y Moel. The towns in the valley had long rows of terraced houses and did not look quite at home in the narrow confines of the steep-sided valley. They existed there because of what was below them in the earth.

On arrival in Nant y Moel the first challenge was to find the pre-arranged lodging at the house of an established miner in the valley. The income provided by a Cardiganshire lodger would be a welcome supplement to the household, and the newcomer would be made welcome. The other priority was to locate the chapel where each miner knew there was a good chance of meeting other neighbours from home.

Life in a mining valley community would have been an eye-opener to John and his fellow lead miners. Many of the terraced houses they lodged in had electricity, and the simple task of flicking a switch on the wall for light, as well as the ability to cook on an electric stove without the need for a fire, was a revelation – as was the ability to turn on a tap in the scullery to access water without the need to take a bucket to the village pump. There was also a plentiful supply of coal to keep the household fire going almost all day. Pavements, street lighting and the close proximity of schools for children were other noticeable advantages compared to the hardship of rural life back home, where there was no lighting and the roads were often dusty in dry weather and puddled with water after rainfall. His children at Salem had a long walk up the hill to Penrhiw and down a steep track to arrive at their school in Trefeurig.

The community in which John Davies found himself was friendly and warm, despite the tough toil of employment. Coal miners were mild-mannered, educated men, with intellectual interests as well as hobbies which included pigeon racing, whippet racing, carpentry, music and singing. Many had been underground since the age of fourteen, and their families existed on low wages.

John Davies would have left for his first day underground, tramping down the streets of terraced houses to the pithead which was dominated by the pit wheels. The wheels would lower men down into the earth and bring up the coal. Before going down he and his fellow miners would be checked and asked to hand over any matches or cigarettes, and they would collect their safety lamps. While waiting for the cage, they could hear its noise and when it arrived on the surface it would release a group of blackened men after their seven and a half hours of hard work, thinking only of a bath and food.

On arrival at the bottom of the shaft, they would walk along the tunnel or roadway for a distance of almost a mile, which was far greater than what John Davies had been used to when underground at Bwlchglas. The roadway was over ten foot high in places, yet would then become lower, requiring them to stoop. There would be a noise and a shout of 'journey' as a row of twenty-five drams full of coal passed, pulled by a steel cable. There were side passages where smaller drams were pulled by pit horses which lived underground and were well cared for.

Once at the coalface they would set to work for another

seven hours. Work at the face was warm and dry compared to the damp and cold at Bwlchglas, but the air was full of black dust. The dust covered the skin and lined the nose and could be tasted with each breath. Work often meant crawling through a forest of pit props, and working hunched over or on knees with barely enough room to wield a pick or shovel. At the end of the shift it was back on the roadway to the shaft and the cage, and ascent to daylight and fresh air and a weary walk back to the terraced lodgings.

In that house, just as it was back home in Salem, the miner's wife was responsible for making sure the rent was paid, the budget to feed and clothe the family, and to have hot water and food ready for when her husband came home. John Davies and his landlord miner would wash their hands and arms, and then sit down for their hot food, which was usually a stew of some sort. After eating, a large galvanised tin bath would be placed in front of the fire and hot water added. They would then bathe and change their clothes. Conversation would continue in a normal and polite way, irrespective of the state of undress.

John Davies would walk up the steep hillsides to look down on the valley, and in the mist evening cloud would hang like grey smoke over the terraces. He'd think of the contrasting greenery of home and his family.

He would attend chapel twice on Sunday, with each visit interspersed with the family lunch. There would be no talk of work in respect for the Sabbath. He had also joined the Miners' Federation, which was affectionately known as the Fed.

This was the pattern of life, day-in day-out and week-in week-out until it was a holiday period when he could prepare and look forward to the journey home. He'd travel down the valley to Bridgend to catch a train, with his wicker basket over one shoulder containing his clothes and other personal possessions. The train would take him to Carmarthen, and then on to Aberystwyth from where he would travel to Bow Street. He'd walk from the little station, over the hill and down to Brogynin and up to Salem, enjoying the quiet of the hills and the familiar green landscape.

Their son Elfed was born in September 1919, nine months after his father's last lengthy visit home for Christmas 1918. The Great War had ended and the process of releasing conscientious objectors had begun. Nancy Astor became the first female Member of Parliament in the House of Commons. John Davies would return home at Easter, and again in the last week of July and first week of August which was the traditional miners' holiday in the south Wales coalfield.

When in Nant y Moel he sent money home every week to ensure that Marged Ann had enough to look after the growing family. He avoided the temptation of drink, as it would deplete the amount of money he could send home in the registered envelope each week.

By December 1921 another daughter, Megan, was born, nine months after his visit home for Easter that year. However, his stay at home this time was prolonged, and he remained without an income for three months. The miners had gone on strike due to the imposition of reduced wages and increased hours of work by the coalition government of Lloyd George. The industry was becoming out of date, and workers like John Davies were cutting coal with pickaxes. Only a fifth of coal was being cut by machine, but mine owners refused to modernise and attempted to cut wages and increase working hours. Miners had called a strike on 31 March 1921 and by 3 April coal rationing had already begun. The strike lasted until 28 June. It failed however, due to the fragmentation of what became known as the Triple Alliance, and the refusal of the National Union of Railwaymen and National Transport Workers' Federation to support the miners. The miners were forced back to work with a wage cut, and John returned to Nant y Moel. Under the strong guidance and incredible financial management of Marged Ann, the family managed to survive. By 1925 the mine owners attempted to cut wages and increase hours again, and the reconciled Triple Alliance threatened a General Strike on 31 July, but Prime Minister Stanley Baldwin and his government intervened and paid a subsidy to prop up wages. By 1926 there were again plans to cut wages and increase hours, and working relationships deteriorated and a General Strike was called which lasted nine days from 4 to 13 May, but the trade unions had to give up in defeat.

John Davies continued to work underground and send money home. The strain on the small household had lessened, as Sal, the eldest daughter, had now left home and gone into domestic service at Caergywydd farm near Bow Street. Lizzie was similarly employed with Edwards the tailors at Penygroes near Llandre. Glenys was studying at Ardwyn County School and staying with a relative in Aberystwyth during the week. Beattie had found work looking after the children of Tyngwndwn farm in Cwmerfyn.

However, things were getting worse on a national level. John Davies came home in early 1930 with a serious chest problem from the effects of dust and mining. He sought

work closer to home but the Great Depression limited opportunities to find work and unemployment was high in the locality, so he was on the dole. By 1932, 36.5 per cent of the working population of Wales was unemployed, and in the south Wales Valleys as many as 75 per cent were out of work. Cardiganshire County Council tried to help local workers by creating low

Marged Ann and John David Davies, outside their home at Salem paid work opportunities in road widening and improvement schemes, and this low-paid work helped the family to survive. Glenys left home to go and train as a nurse in Walsall at the age of eighteen.

Their son Elfed was now fourteen and was considered a bright boy. However, the lack of family income meant that they could not afford to send him to secondary school. He had to seek employment, but was determined not to be drawn into agricultural work. He was lucky to find a vacancy at W.H. Jones, an ironmonger and china store in Great Darkgate Street, Aberystwyth, where he would be apprenticed under the guidance of H.G. Snell, who had been invalided out of naval service as a radio operator.

Elfed's friend Ifor James from Salem also managed to find a job, and they both arranged to live in town and moved in with Elfed's sister Beattie who was in domestic service with a family living in Grey's Inn Road. The family agreed for them to stay in return for a rent of five shillings, which wiped out Elfed's weekly earnings. Any spending money would be entirely dependent on tips earned from customers at the shop, or on delivery of goods to them. Elfed and Ifor bought two second-hand bicycles, and would cycle home to Salem on Saturday night after the shop had closed and be back by 8.30 on Monday morning. By 1935 Elfed had saved enough money to buy a new bicycle with a bell and carrier for £3 18s. from J.B. Morgan garage in Penrhyncoch, which would make the journey easier. He enjoyed the work and living in town, and met a young woman, Lena James, who lived in Edge Hill Road, and they spent quite a lot of time together, often accompanied by her sister Jane.

Illness was rife in the countryside, with diphtheria, measles and scarlet fever taking its toll on many families. Elfed's family continued to suffer its share of hardship. His sister Megan became lame through a dog bite which went untreated as they could not afford to pay a doctor. His other sister Hannah Mary contracted TB and spent some considerable time at Llangwyfan sanatorium near Denbigh in north Wales. She returned home to Salem but died from the disease in 1937, and was buried in the cemetery of Salem chapel.

There was considerable disquiet and unease in Europe due to the increased hostility of Germany. By the summer of 1939 Elfed, like all other young men in the country, had to register for National Service. War was declared by Prime Minister Neville Chamberlain on 3 September 1939, and by 15 December Elfed was required to report to the Royal Artillery Anti-Aircraft camp at Oswestry to begin his wartime service. This would take him to North Africa and Italy.

Despite the demands of enlisted life, Elfed and Lena's relationship continued to flourish. In July 1941 she had enlisted in the Women's Auxiliary Air Force (WAAF) and travelled to Blackpool for her basic training. Elfed, who had been on leave, managed to visit her by arranging a detour on his way home to Salem. In March 1942 he received a telegram informing him that his father was gravely ill. He travelled home but John Davies had died of pneumonia on 8 March. There had to be an autopsy, as he had been a miner. Without doubt, his mining employment was a contributory to his death, but for a long-serving miner he had lived to the reasonable age of sixty-five and was buried in the cemetery at Salem chapel alongside his daughter Hannah Mary. His tombstone inscription reads: *Mi a ymdrechais ymdrech deg* (I fought a fair fight).

As they were home on compassionate leave, and Elfed knew he was soon to be posted overseas, they made the hasty decision to get married quietly at Tabernacle chapel in Aberystwyth, with the Rev. J Meredith officiating. There was little choice for a best man and guests, as all his friends and contemporaries were away on active service. A modest reception was held at Lena's home in Edge Hill Road, and they had a one-night honeymoon in Shrewsbury on their return to their respective units.

They both returned to active service and, by 8 May 1942, Elfed was on a convoy ship heading towards the Red Sea to begin more than a year of service in Egypt. In August 1943 he would be in Italy and would not meet up with Lena until he had been allocated a month's leave in August 1945.

His new wife met him at the station in Aberystwyth and they spent a hectic time together visiting both families as well attending various welcome home functions. Lena was released from service on 7 August 1945 and returned to civilian life to take up her old employment at Mrs Dewar's shop in North Parade, Aberystwyth. Elfed had to return to Italy to see out the remainder of the war and was finally demobbed in April 1946.

In June 1946 their son Brian was born, and in November Lena's sister Jane gave birth to their daughter Valerie. Lena's parents, William Richard and Lizzie James, purchased a new house on the opposite side of the street: a terraced redbrick property, with three bedrooms and two small living rooms downstairs, a small kitchen with cold running water, and an outside flush toilet next to a small outhouse which housed a water boiler for use on washing days. Considerable work was done by William Richard, Elfed, and his brother-in-law Owen to install electricity and completely refurbish and redecorate the property, despite the post-war shortages and rationing. They were all able to move in by May. The house had spectacular panoramic views of Aberystwyth town from the far end of the small narrow garden, extending from Tan-y-Bwlch, the harbour, across to the castle and the promenade, to Constitution Hill on the north side.

In May 1947 Lena gave birth to their second son, Nigel, and the small house became even more crowded with a total of nine people living in it. Somehow they managed, despite the post-war austerity. Some relief from the crowding came when Elfed and Lena and the boys travelled by bus to Penrhyncoch at weekends. They would then walk to Salem to stay with his mother Marged Ann.

Elfed secured a job as a wireless mechanic at an ironmongers' shop, Humphrey Jones & Sons. They were the new owners of the shop in Great Darkgate Street where Elfed had worked before enlisting. Post-war, the shops now closed at 5.30 p.m., and there was half-day closing on Wednesdays. Public transport had also improved, with a regular daily bus service from Penrhyncoch to Aberystwyth departing at around 8 a.m. to allow people to travel to work. This enabled the family to spend some days in Salem, relieving the strain on the house in town.

As a result of an improvement in the water supply in Aberystwyth and some of the surrounding villages, the Rural District Council undertook some house building. Twelve new houses were built below Garth in upper Penrhyncoch. Priority was given to newly married ex-servicemen, and Elfed and Lena were fortunate to be allocated the rental of a new house into which they moved in 1949. There was still no electricity supply available, despite the fact that the wiring had been installed. Cooking was done by a coal-fired combination oven which was located in the rear living room. This provided water heating as well as cooking hot food. Lighting was by means of paraffin oil lamps, and morning tea was achieved by a primus stove prior to lighting the fire later in the morning. It was a large comfortable semi-detached house, with three bedrooms and a bathroom upstairs, and two living rooms with a small scullery kitchen downstairs, alongside which was a walk-in pantry.

An outside shed and coalhouse stood beside a generous sized garden, which would provide fresh produce for the family.

In January 1952 Elfed embarked on a new employment opportunity. He successfully applied for a post as a van salesman with the rapidly expanding Ever Ready Battery Company. This improved his earning capacity and also enabled him to apply some of the skills he had learnt during his Army service. He would work across the county of Cardiganshire, and travel to and from home in the van provided by the company which he kept secured at Tŷ Mawr Garage in Penrhyncoch.

The author's father, Elfed, and the Ever Ready van

My father would frequently reflect on his childhood in Salem, and would fondly recall his life in this small hamlet. He eloquently wrote an article for a local paper remembering the simple village life.

My Birthplace (Bro fy Mebyd)

My birthplace covers a wide circle which includes part of Penrhyncoch, up to Penrhiw, Llwynprysg, Banc y Darren, Cwmerfyn, Cwmsymlog and Trefeurig. It was the circle surrounding the school at Trefeurig. The village I was raised in was called Salem, Coedgyffudd, about a large field away from Brogynin, the home of Dafydd ap Gwilym. In the village there is a fairly large chapel, a shop and two wells which never went dry. I can assure you that I drank gallons of water from them, as there was no mention of pop in those days.

This place was very cold in the winter but in the summer there was no better place, with the cuckoo singing every day and all day for about two months. There are three roads which you could take to go to Salem, through Garth in Penrhyncoch and up past Brogynin, or up to Penrhiw and down to Salem, or straight up after turning at Brynhyfryd. If the wind blew from a direction and it was snowing, then the lane would be shut for weeks but that did not happen often.

I was one of seven children, six girls and one boy, but I lost one sister when she was twenty years of age. TB was very bad in Cardiganshire in the Thirties. It was not always an advantage being the eldest, as there was some responsibility for the eldest to set an example. Work for the men of the village in those times was lead mining at such places as Bwlchglas and Camdwr. Walking to and fro, working in very unhealthy places for little money. Water came in on their heads and they had to be in wet places for hours. They worked shifts; one shift would start at two o'clock in the afternoon and coming out at around eleven at night, then walking home over the hills.

It froze hard on the tops in the winter, and by the time they reached Salem their clothes had frozen on them and they would have to stand in front of the large fire to thaw and get out of them. If you go to the cemeteries you could see that there were many in their graves before reaching forty years of age, which is no surprise.

The lead mines closed and many of the men went to work in the South in the coal mines, but as William Jones told me 'it is a strange place down south'! I heard them say that one of the strangest things was to have a bath in a pan in front of the fire and the woman next door coming in to borrow some sugar or flour. They got used to this as the wages were quite good, but they had to leave their families and only go home for a week at Christmas, Easter and two weeks in August. That was the only time they were home in the year unless there was a strike. It was no surprise that many of us children had their birthdays at the same time.

Coming home from the South was a bit of a problem, no buses or cars and dependent only on the trains. Changing at Carmarthen and a slow journey on the train. Many would have been working in the mine overnight, then standing on their feet in the corridor probably until they arrived at Aberystwyth. Then a train from Aberystwyth to Bow Street or Llandre. Then walking, some to Penrhyn and Trefeurig or Talybont and other places. There was no telling when they would arrive and we children would look at the clock and put a kettle on the fire. At last they would arrive at the Easter holiday or in August and some of the younger children would not recognise their fathers. The mothers with a baby wrapped in a shawl would walk to meet them.

They never carried a case, only a basket with a strap around it to secure it to their backs. It was a sad occasion to see them leave after being home. Then a letter would arrive with money every Monday morning. Things were bad during a strike, it was good to have the fathers home of course but there was not a penny coming in. There was no dole or anything else. I remember my father and many other fathers home for four months during a strike, the mothers getting together and crying as it was so hard. However, I do not remember that we had to go to school without breakfast or in bare feet. I was too young in those days to realise, but now I can look back and I believe that the age of miracles had not ended and nothing makes me angrier than to hear people today saying things are hard. Poor things.

There was an old lady working as a nurse in those days, helping women when having their babies. But in 1912 a District Nurse came to the area for the first time and it was expected that all would support her. I cannot remember how much the charge was, but I remember my youngest sister being born on the 28th of December 1921 when I was two, and understanding things but not as much as the children today.

When important things happen they always happen at night. I remember my father walking with a lantern down past Tynpynfarch, up Cwrt Lane and down past Tynrhos to call on Nurse Rees. If such a thing happened today there would only be a need to phone or jump in a car, but not in those days. After alerting the nurse my father called on Mr Goronwy Rees. He, poor soul, would not be happy, but out came the pony and trap to take Nurse Rees up to Salem. My father and his lantern were also in the trap. You can imagine the clippety-clop through Bow Street in the dead of night, past Gogerddan and up through Penrhyn. The men, I'm certain, would need to walk past Horeb chapel where the road was steep. Up to Salem and all being well, as luckily there was no snow that night. After that Nurse Rees would come up to Salem every day for about ten days on her bike. She was a very nice lady, not only looking after the baby but keeping an eye on all of us as well.

Well, within about two days, my father's holiday would be at an end and he would need to go back to work. Seven of us would expect food, and an aunt would come over to look after us, and I would need to roll up my sleeves as well. They were hard times, but everyone was happy in a close community in which everyone helped each other.

We were allowed to sow around two rows of potatoes in a field owned by the farmer at Coedgruffydd. There was no charge for sowing the potatoes, but the women were expected to help with the harvest and I, like many others, would be expected to look after the younger children. I have a memory of taking the youngest to the corn field and the mum sitting on a corn sheaf and breastfeeding the baby after which I would take it for a walk in some old pram.

There was plenty of room for play in the summer, and plenty of wild flowers growing. We played little weddings frequently, and collected blackberries or winberries from the hedgerows. Life was full of entertainment. Plenty of food, with the occasional rabbit available then, with a good garden full of vegetables and killing a pig each autumn. I remember going to gather stones in fields for threepence and free tea in the field.

There were very amusing characters in the locality. I recall an old lady who lived next door to us. Lisa Williams was her name, a short, stocky woman. She had purchased wallpaper for the kitchen; there was no lounge or dining room then. The paper cost twopence a roll, and Lisa would get up early in the spring to work in the kitchen and hang a length behind the kitchen door. She then called on my mother who said, 'You've started papering Lisa?' 'Yes,' she'd replied, 'but I don't like it.' Mother replied, 'it will be all right when it's finished.' 'No,' said Lisa, 'I will take it back.'

She sat in the chair and put her boots on, then her coat, and walked six or seven miles to Aberystwyth from Salem. Nowadays we would make a cup of tea before going in the car, but there was no chance of a lift then.

I also remember Edward Hughes, Isaac Hughes Afallon Deg's father, bringing butter from Llawrcwm Fawr farm to the shop in Salem. Walking, of course, with his stick as he was more than a little lame. If we saw him we would run home and then to the shop to buy the butter, as the Llawrcwm Fawr butter was very nice. I don't know why he couldn't deliver the butter to the house, but that was the way it was. He came with two sheepdogs and, after selling the butter, he would come over to Tom and Ann Edwards to have tea. They were great friends, but the dogs would not be allowed in the house as Ann was very house-proud. The dogs would be at the door and we children would go to them and sing, and the dogs would sing with us. Edwards in the house would hear the concert and say, 'Good God, Ann, if I don't go to heaven the dogs surely will after all that singing.'

To go back to Tom and Ann, two dear old characters who were childless but who were very kind to us children. Sometimes, when Mother would be away from home, which didn't happen often, Ann would ask if she could do anything to help and would care for us. She would ask if we needed anything, and we would ask for bread as there would always be a large loaf, having been baked in the wall oven. Ann was very clean and would always be polishing. They had oilcloth on the stairs with brass rods holding it in place. Polishing oilcloth is a dangerous thing, and the stairs were opposite to the front door. One morning in the summer, Tom came down the stairs in his stockinged feet and slipped with both feet and flew down the stairs to land on the main road. There was no danger as there were no cars in the village, only the occasional bike or sheep passing.

I recall two other innocents living in Lluest Fach y Mynydd, a little further up from Cwmsymlog, known as Dafydd and Mari Ann. They came down to the shop and Dafydd always kept his hat on with string. One night, they came down to Trefeurig and forgot to bring their lantern, as they intended to be back before dark. They borrowed a torch from a friend in order to walk home in the dark. After arriving home they started blowing at the torch in an attempt to put out the light, but they failed. They then tried with the bellows and, after a long attempt, the torch began to dim but still was not out. They wanted to go to bed but were afraid to leave the light on all night in case of fire, so they placed the torch in a bucket of water and slept peacefully.

Our school was at Trefeurig, which was not a long way from home, up a steep hill to Penrhiw and then down a bank to the school. We and a group of children from Llwynprysg and Penrhiw would meet on the hill before going down to the school. The hill was bare and cold when the wind blew and the rain fell. We could see the kids from Banc y Darren coming down on the other side. There were around a hundred at the school at that time, as there were some large families in the area.

We had some good teachers. Dan Jones was a schoolmaster from Bronant, but I was only with him for six month. He had a good influence on the area, and he brought in a lot of Welsh.

I would spend time going home from school dreaming, or

looking for nests or following butterflies in the bracken to arrive
home for a telling-off. We would spend a lot of time in the summer
collecting firewood, cutting bracken to put as bedding for the
pig, and collecting nuts and other things. Nearly everyone kept a
pig in the village and they would be killed in the autumn. We, as
children, would cry when the pig was killed but I often think of the
bit of bacon we had in those times.

We attended chapel regularly and we had a great Sunday
school. There was a service in the morning, then the minister
would go up to Cwmerfyn for an afternoon service, and walk the
whole way. Then there was Sunday school in the afternoon and a
prayer meeting in the evening, with plenty attending to support the
service. Once a month we would have a meeting for recitation and
singing, for which we would learn verses to recite and twosomes or
foursomes for the singing. When the sermon was in the afternoon
the mothers would take their babies to the chapel in a shawl, and
if they began to cry they would go out to the porch to rock them
to sleep and then return to the chapel. Sermons could be long in
those days.

William Magor had a brake which went to Aberystwyth every
Monday – that was a turn-out. The brake was like a half-barrel
pulled by one horse, and everyone needed to get out and walk on
the hills despite rain or wind. William Morgan lived in Penyberth
near Penrhyncoch, and it was only his brake which went up to
Salem despite there being others in the locality.

I walked back from Aberystwyth many times, and there were
rumours of a ghost on the road and a White Lady near Gogerddan,
but I never saw anything. The most frightening thing would be to
hear a sheep coughing in the hedge.

Well that broadly and untidily is the story of my homeland
and, although I have been away for some years, I still enjoy going
back, especially in the summer or the winter. Up to Penrhiw to
look down at my old school in Trefeurig, and then down through
Salem. But now most of the residents are strangers to me. I can't
knock on a door and say, 'How are you?' That is how it now is, the
world has changed.

The war had brought about many transformations to old
established orders and practices. Changes continued in the

countryside. In 1952 the Gogerddan mansion and remaining 3,839-acre estate belonging to Sir Pryse Loveden Saunders-Pryse was sold. Parts of the estate had already been sold to meet debts and family commitments. The mansion and estate, which had a long history going back to the 16th century of gentrified ownership and land management, was sold to the University of Wales to establish a plant breeding station. The Pryse dynasty had owned a large number of farms and had a long history of mining in the locality too. The link died with the death of Saunders-Pryse and the baronetcy became extinct. However, the new post-war challenges and changes brought the community together, and the foundation for a new village hall was cut in a parcel of land adjacent to the new Maes Seilo houses.

Lena gave birth to their daughter Avril in November 1954 and Elfed's secure employment in the rapidly changing post-war world allowed the family to flourish in this compassionate rural community.

The changes brought about by war would continue, but through recognising the past and appreciating the privileges of the present, the new generation would look forward to an optimistic future.

4

Train

Isaac Jenkins
(1883–1966)

ISAAC JENKINS LIVED in Gamlyn Cottage in the hamlet of Aberffrwd in the quiet, remote valley of Cwm Rheidol. He'd been born at Lluest y Parc farm in the nearby village of Ysumtuen, and earned his early living as a farm worker, working with his father who was not keen for his son to go underground to work in the local lead mines. Ystumtuen was a scattered hamlet located in a lead mining area which had been mined since the late 17th century. It had a varied history of profit and success. The workforce in the mid to late 19th century included a large number of Cornishmen who had constructed the centrally located Ebenezer chapel. But by 1894 a large number of the workers, both above and below ground, were laid off due to the decline in demand for local lead as the result of imported ores.

Isaac had met and married his wife Catherine Ellen Williams and moved from Ystumtuen down to the valley and into the comfort of a small cottage which stood beside the narrow road which tortuously connected the community and the hillside above. The little house stood next to a green meadow, which led down to the banks of the River Rheidol. He had been well educated at school in Ystumtuen and by his father, as well as attending Sunday school regularly. His father had ensured that there were other positive influences on his development. The family home had a variety of books which he had eagerly read, and he ensured that his own home in the valley would also have a wealth of reading material.

The cosy living room had a large fireplace, and a bread oven set higher on the left. It would be used once a week to bake bread for the family. A small scullery-cum-kitchen overlooked the field and the river at the rear of the house. Three upstairs bedrooms were accessed by a narrow stairwell. Two of the bedrooms overlooked the field, while the third had a view of the road. At the rear of the house was a small garden which grew vegetables, including good crops of potatoes, peas, beans and also rhubarb. The edges of the plot were lined by gooseberry and blackcurrant bushes. The garden would be dug over in the autumn and well manured ready for early spring sowing. In one corner stood a small wooden shed which housed a few chickens which would be released to wander the field each morning, and then securely locked up at night out of the way of the numerous foxes that lived in the densely wooded valley.

The steep, tree-lined sides of the Cwm Rheidol valley and the tranquillity of the flowing river in the valley floor created a rural serenity, complementing the relative isolation of the valley. However, there was industrial activity in its confines, and the head of the valley contained a number of mine workings, some of which had been there since the 18th century. The mines had been owned and worked by numerous speculators, and they had seen several rises and falls in their fortunes.

However, they had provided work to many who lived here, as well as attracting others to come and work there, despite the difficult access and lack of transport.

Mining operations were intermittent however, and subject

to the whims of local or distant investors. Mining activity left scars of barren spoil, and poisoned the soil with zincblende, polluting some of the water.

The close-knit valley community kept a small-holding lifestyle in order to support their families, as wages from mining could be unreliable. Many of those living in the scattered community were interrelated, and some of the children would be employed at the mines as surface workers, to supplement meagre family incomes. This often involved doing ore dressing work, which could have a detrimental effect on their future health, despite the fact that at some mines the ore dressing areas were covered to protect them from the rain.

Sundays required attendance at the Methodist chapel located a short distance along the road from Gamlyn Cottage. The chapel had a long-standing influence on life in the secluded community, having originally been built in the 18th century and rebuilt in 1884. The plastered stone exterior, with a slate roof, had a simple austere interior with an imposing pulpit which was accessed by a pair of curved staircases with timber balustrades. The plain painted walls were well lit by round headed windows, and the high ceiling had an ornate central plaster rosette which looked down on the rows of boxed pews.

Here the scattered community would congregate to attend services twice on Sundays, as well as supporting a Sunday school for the children. It was a vital meeting point where local gossip and information would be exchanged, as well as ensuring that the extended family in the valley were made aware of the health of their distant relatives. This regular information exchange was important in a community where everyone knew everyone else and nothing could be hidden for long. The gossip provided unity and cohesion, as well as support in times of difficulty, so that each one in their *milltir sgwâr* (square mile) felt part of their *bro* (area).

In 1901 work had begun on the construction of a narrow-gauge railway to link Aberystwyth to Devil's Bridge. Its purpose was to transport lead ore more efficiently to the town's port, and this despite the declining fortunes of the local mining industry. The railway line was completed by 1902, and, as already noted, profited from some of the redundant engines and rolling stock from the Plynlimon and Hafan Tramway which had closed in 1899. The trains delivered twenty tons of ore per day to the harbour's quay in 1903. But, the lead mining industry was in serious trouble due to cheap imports. The railway climbed 700 feet over twelve miles from Aberystwyth to Devil's Bridge.

There was also demand for timber which could be transported down the valley, but the sharp curves of the line made this a challenging undertaking. However, the line gave Isaac an opportunity for employment, and he joined one of the gangs who were employed in maintaining the line and guaranteeing its safe operation over the many sharp bends. The work also enabled him to stay in the valley and his local knowledge endeared him to his fellow workers.

As the goods and ore traffic dried up, the railway became a more profitable venture as a passenger system and especially as a tourist attraction. Isaac had the benefit of a regular and dependable income, which was in contrast to those who still hoped for a revival of the mining industry in the valley.

Isaac's passion for the work and his love for his home area was reflected in a descriptive piece he wrote on the railway line and the valley:

A description of the Rheidol Valley by Isaac Jenkins, Ganger
For those who are staying in Aberystwyth and wish to see the beauties of nature, one of the remarkable valleys in the district, or even in Wales, is the Vale of Rheidol. Here nature is always to be seen at its best. Nature, which never goes on strike, a member of a union without strikes on the programme.

Through the valley runs the Vale of Rheidol Railway, a narrow gauge of 1' 11.5" width which runs on the hillsides in a position that enables the passengers to have a grandstand view of the valley below, gems of views which can only be seen from the *trên bach*.

The train starts its journey to the famous Devil's Bridge from alongside the parking ground, across the Smithfield, on to a sharp curve under the Carmarthen Branch, alongside a plot of gardens where the railway employees grow their veggies, and on to Plascrug, the palace of the Welsh Princes. Then it runs parallel to the main line on to Llanbadarn, where an ancient and historical church is to be seen. The only place of worship in the whole valley until the year of 1756, when the Calvinistic Methodist church opened in Aberffrwd, is further on. From Llanbadarn the railway keeps to the south over the Rheidol river bridge on to Glanrafon halt, coming in touch with the river now and again. Very soon you may glimpse the Lovesgrove Mansion, then again on the left with the ground gradually rising in the direction of Gogerddan, the residence of the well-known Pryses.

This stretch of land forms a beautiful setting for the mansion, with its clumps of trees here and there amongst the meadows. Before you, from here towards the north, but at a far distance over meadows and cornfields, you can see the hill peaks here and there. The little train puffs along through the fields, everything is so pleasant on both sides. On the right hand side is Pwllcenawon, where the Rev. gentleman, Mr Lewis Edwards, was born in the year of 1809, later to become the principal of Bala College. Also, a well-known musician was born in the neighbourhood at Troedrhiwfelen, called John Roberts, better known through Wales as Ieuan Gwyllt. It was he who started our Welsh singing festivals.

Within half a mile you will have come to Capel Bangor station, a distance of four and a half miles from Aberystwyth, where you will notice you are dead level with the River Rheidol. At a little distance on the left is Capel Bangor and Penllwyn village. Another valley opens here, called Melindwr, where the main road runs to Goginan, Ponterwyd, Llanidloes and out to the Midlands. The narrow gauge follows the Rheidol. Very soon, on leaving Capel Bangor station, the little engine is making a dash for an incline of 1 in 50 through a grove of trees with their branches leaping over the train.

Suddenly you come into the open again, onto one of the most remarkable spots you ever saw. The cornfields and the meadows are below you, with their hedges running zigzag back and fro, and the hillsides covered with trees in their green garments on. A dingle here and there, giving way to a ravine now and again,

107

beautifully clad with trees. A little brook runs down, murmuring and singing over the rocks and between the stones, and in sight of the Rheidol, flowing in the centre, turning and twisting forming the letter S or V in different places. The train arrives at Aberffrwd, an altitude of 200 feet. While the little engine is having three or four minutes to rest its engines at this station, you gaze around to see standing majestic in front of you the Silver Hill, which had its name from the lead and minerals, the scene of the labours of our forefathers for many years. Below is Aberffrwd Chapel, where they used to gather on Sunday morning to worship. On the right of the hill is a dingle running towards the Devil's Bridge main road. The train is following the Rheidol on to a mining area, with some rugged heaps here and there. Soon after leaving Aberffrwd the engine makes a bold attack with fine smoke and fury on to a continuous incline of 1 in 50 over some very sharp curves and S curves. You may get a glimpse of the Glanaber Falls.

Suddenly, around a corner, you will see the thrilling sight of the stag well up on the hill, a realistic nature carving of the wild stag itself. Suddenly, the Rheidol Falls come into view as the train puffs strongly, a beautiful cascade faces you, with the water tumbling down rocks gleaming as white as snow as you spin around a sharp bend on to a curve called the Horseshoe Bend. On the right there used to be a lead mine, and you will also notice a very deep and rugged ravine with its dangerous precipice. On the other side you will notice some scattered houses and little farms, with their fields coming onto the brink of the hilltops. Further on a little chapel stands a solitary guard over a few tombstones. It makes you wonder whether it will not be forgotten in the day of resurrection.

Before the railway the remoteness of the valley had given rise to a number of self-sufficient small-holdings. These kept chickens, geese or a pig, and a vegetable garden to provide food for the family. And when illness struck the inhabitants, they would use a traditional cure rather than pay for a doctor from distant Aberystwyth. At the head of the valley lived a well-respected woman named Elinor Mason. She collected herbs from the hedgerows and also grew her own in her garden. She would supply herbal cures for a number of ailments for a small payment or a bartered exchange of goods or food.

Other more obscure concerns and anxieties would be dealt with by a *dyn hysbys* (soothsayer) or *consuriwr* (cunning man). These wise men would be consulted to deal with shadowy issues, ranging from sour milk to depression. The cures would normally involve a spell or charm, which would be secretly accepted by the anxious sufferer, as there was disapproval of such remedies by chapel leaders who claimed it was the work of the devil.

Folk magic had a long tradition in the valley and wider area, with well-known providers such as Evan Griffiths, Llangurig, and Dicky Davies, Fagwyr Fawr, Ponterwyd. Often, in desperation, people would arrange to visit these men who, in turn, would pay a return visit to the valley to support the spell or cure which had been applied. Drinking spring water with saffron was said to cure repeated intoxication, and preventing unchaste thoughts was cured by eating rue herb in the morning. They would also use miracle stones or *maen magl* to cure a variety of ailments, especially those involving difficulties with sight. A *dyn hysbys* would take into account the fears and superstitions of a locality. They often claimed to be able to foresee the future, but they were mainly consulted for their traditional medical knowledge and could offer treatments for men, women, children and animals. They had an esoteric folk knowledge of herbal cures and remedies, which had been passed down for several generations.

The railway changed the lives of most of those living in Cwm Rheidol, Aberffrwd and Devil's Bridge. Before the train their journey to Aberystwyth was long and tedious, and would only be done about once a week, usually on a Monday.

The railway thrived, with several services a day going up and down the valley. By 1905 it was agreed that the train could carry mail, and the line was paid £120 per annum for the service which required at least one mail train to run each day.

Open-air carriages were introduced from 1909, and by 1910 there was the option of first-class accommodation.

Isaac Jenkins would work at Aberffrwd station at busy times to issue and collect tickets, and would often be there from 6 a.m. to 9 p.m., but in between he worked on the line or would be busy digging out a recess to house a new water tank at the little station.

An aerial ropeway with two steel cables was constructed at Rhiwfron Halt. It spanned across the valley to the mine at Cwmrheidol. The cables had two iron buckets, and were operated by two men in order to transport ore from the mine on the opposite side of the valley across to the small railway halt. Here the ore would be discharged into wagons on a small siding, ready to be transported down to Aberystwyth and exported.

On one occasion, a woman who lived at Llain Cottage, close to the mine, returned on the train from market at Aberystwyth, having purchased a small pig. She complained poignantly to the ropeway operators about the difficult journey she had to undertake down the steep sides of the valley, across a footbridge and up the other side with her pig in tow. The men took pity on her and placed some old sacks on the floor of the ore bucket, and she was then transported with her pig across the valley, with just a short walk home afterwards!

Ore traffic on the line declined. By 1913 it was down to twenty-five tons a month. The ore had primarily come from the mines at Cwmystwyth, Frongoch and Cwmrheidol, but some ore was still transported to the train by pack ponies coming from Ystumtuen and from as far as Dyffryn Castell.

During the war years of 1914 to 1918 the line did well, as there was a boom in demand for timber for pit props. The line made an income of £4,000 some months, but passenger services declined. By 1921 only the Cwmystwyth mine was providing ore, but the railway line still made some money transporting produce from farms and gardens in the valley to market in

Aberystwyth, as well as carrying building materials and fertiliser etc., up the valley to the various stations. Goods traffic then also declined, with some farms using small lorries.

There was some local opposition to the railway line by some residents in the valley, but Isaac's local links helped pacify some. Occasionally, the owners of the railway would have to pay out for the sheep and lambs killed when they wandered on to the track.

The aerial ropeway stopped in 1925 due to a decline in the production of ore. Regular local passenger traffic also reduced, as many in the valley were now using a bus service.

On one occasion, in 1929, Isaac found that the timber bridge over an old mineshaft, about nine and a half miles up the track from Aberystwyth, had collapsed. There was a gaping hole below the rails. He walked home to Gamlyn Cottage and collected the washing line, but the depth of the void could not be plumbed with the washing line. He set about constructing a new bridge with sections of old rail, which he slabbed over with rocks and oak saplings felled from nearby hedges. The repair was made good, and lasted for nearly fifty years until 1979.

His family expanded and their third son, Arthur, eventually joined him on *Lein Fach* as a ganger. His son Owen, born 4 April 1923, also took an interest in the railway line and gained employment as a fireman on the footplate alongside the train driver. In 1938 the train operating company caused considerable concern in the valley by attempting to enforce a regulation that no working instructions were to be given in Welsh.

Isaac Jenkins kept notes on aspects of his work, especially the detailed information on the intricacies of the convoluted track which clung to the steep side of the valley.

His pride in his job and his conscientious approach to his work is well illustrated in a note he wrote:

A Spotlight from the Diary of a Railway Ganger

Thousands of tourists and holidaymakers visit the seaside resort of Aberystwyth each year. It appears that there is a paucity of guides and adverts to the interesting attractions in the various localities. One of the best attractions in the district is the Vale of Rheidol Railway, a narrow gauge which runs from Aberystwyth to the famous Devil's Bridge falls, one of the most remarkable features in the world according to the well-known writer George Borrow.

A name that is always associated with the Vale of Rheidol Railway is Mr James Rees, who was General Manager from its inception until 1915, and to him is due all the praise for popularising and advertising the toy railway, and for the excellent manner in which he conducted the affairs of the *Lein Fach*.

Famous journalists visited the valley and photographers, with their huge cameras, swept the countryside. Spencer Leigh Hughes, of the *Morning Leader*, arrived at Devil's Bridge by an afternoon train, and there was a trolley at the station waiting to take his party down the incline towards Aberffrwd. The party enjoyed a picnic at a spot called the Horseshoe Bend, and then proceeded along to Aberystwyth. The following week articles appeared in the daily papers and in the London journals describing the surrounding district, with the hills and the puppy hills here and there, and the beautiful scenery of the Rheidol Valley: its cascades, dingles and ravines, all of which can only be viewed satisfactorily from the *trên bach* (little train).

In its early days, nine trains used to run each way every day during the summer months, with every coach full to its capacity. Sometimes, two engines were used coupled, puffing along with fire, smoke and fury, with ten coaches behind. The traffic was so heavy that the company was compelled to borrow an engine from the Ffestiniog Railway. Old Dafydd was the constant driver of the Ffestiniog engine, named *Palmento*, and so attached was he to his engine that it was said that he used to sleep on the tender at night.

Beside the summer traffic the railway was paying its way during the winter months when the lead mines were working at full swing. Cwmystwyth, Dyffryn Castell and Ystumtuen mines sent thousands of tons of lead ore to Devil's Bridge to be conveyed to Aberystwyth and on to other destinations. Moreover, the Rheidol mine, at the beginning of the century, was experiencing a busy and flourishing period. It was commonly called De Bal's mine, after the engineer who constructed the aerial cableway worked by electricity to convey ore across the River Rheidol to the railway wagons at Rhiwfron Halt. The capacity of the buckets, which transported the ore, was five cwt. And full containers ascended as the empty buckets descended. The machinery for dressing the ore was worked by a huge turbine fixed opposite the mine by the riverside. Further up the river, a dam had been built to turn a sufficient supply of water to the turbines.

All the machinery, materials and implements necessary for the upkeep of these mines were brought along by the *Lein Fach*. As traffic was expanding so rapidly, the company had to provide timber wagons to carry timber that was coming to Devil's Bridge from the Hafod Estate. It was a great wonder to see such long lengths of timber being conveyed along the winding track of sharp curves and twisting corners on such a small gauge. The railway staff had seen a good many incidents along these sharp curves that they will never forget. I read about the King of Egypt being troubled by a plague of frogs and locusts. We saw a plague of caterpillars preventing the train on its journey, and actually brought it to a standstill!

What a blessing the railway was to the folk of the Rheidol Valley and surrounding countryside. The farmers were supplied with lime and manure for their land, and all provisions for the country merchants were delivered by the railway. The poor miner who lived on the slopes of Pumlumon was able to have a lump of coal to put alongside the peat he dug from the marshy land around him to keep himself and his family warm during snow storms.

Although circumstances and conditions have changed considerably during the last few years, it is most interesting to meditate upon the habits and customs of our forefathers who lived many years ago in the lonely dingles and on the mountain slopes without any communication with the outside world, the nearest town being at least fifteen miles away. The only conveyance from the Rheidol Valley was on a Monday. A farmer from the valley had an old-fashioned market car, something similar to a gypsy tent on wheels. You could not expect to reach Aberystwyth in under two hours, as the horse never ventured to trot until he got under Llanbadarn railway bridge. If you wanted to visit Aberystwyth on any other particular day in those days, you would have to step out and walk it both ways.

There were many instances, like the old lady from Cwmystwyth who regularly walked the sixteen miles to the town carrying a basket containing her small-holding's weekly produce, and returning with a bushel of wheat on her back to grind for making bread for her family in the lonely hills. She used to leave the wheat at the mill on the Mynach riverside between Devil's Bridge and Cwmystwyth. There were about half a dozen such mills along the Rheidol between Aberystwyth and Ponterwyd in those days. Farmers brought their barley and oats to grind, for in those days they lived on their home production. They made bread from barley and wheat and crushed oats in different ways so as to make milk porridge or to mix with buttermilk to make *uwd* (porridge). People lived very simply and cheaply, and a ewe could be bought for five shillings and a lamb for 2/6. Farmers had to go a long way and beg cattle dealers to come and buy surplus stock. Labour was very cheap. A head waggoner on a big farm earned only £20 per year, which was considered a good wage. Somehow, people were living quite happy and comfortable without the slightest idea of looking forward to a brighter horizon and better ways to come.

The coming of *Lein Fach* changed this background to some extent. Many years before the *lein* came through the valley, there was a murmur in the air that it would come in the future. When it did come it brought a new life to the locality, the inhabitants had a better way of communicating with the outside world, and it linked together the more outlying districts.

At the opening ceremony of the *Lein Fach* to passenger traffic, inhabitants over a wide area of many square miles came to Devil's Bridge in traps, on mountain ponies and on foot. This sight became a regular occurrence on hiring fair days, and young folk were to be seen running down the hillside as if they were coming from the wilds and singing and whistling 'a hey ho come to the fair'.

The railway line's manager, Mr James Rees, was able to pay staff wages for three months from the takings on hiring fair days only. There is a great difference between the wages of those days to what they are today. Men maintaining the track were receiving seventeen shillings per week, working from 6 a.m. to 5.30 p.m. Many of them have gone to life beyond the veil or living on their pension.

The *Lein Fach* still runs during the summer months and it

depends entirely now on tourists and visitors. Will it survive and what its fate will be are precarious questions, as scarcely anything has been done to bring its charms to the notice of visitors. The beauty and romance of the Rheidol Valley remains, and can be seen by anyone who ventures along the 1 foot 11½ inch gauge.

As changes took place in Europe, and Germany's aggression resulted in the declaration of war on 3 September 1939, Isaac's son Owen was too young to enlist. He continued with his work until the *Lein Fach*'s activity was suspended. He, Isaac and brother Arthur worked on the maintenance of the line, locomotives and rolling stock.

Owen met a young Aberystwyth woman, Jane James. Jane was a lively woman and, as well as seeing Owen, she would also meet up with other young men in the town. As already noted, her sister Lena would chastise her for her cavalier attitude, and eventually convinced her that the young man from Aberffrwd was a worthy suitor. Lena enlisted in the WAAF. Jane, being younger, was too young to enlist but eventually joined the Women's Land Army. Owen and Jane were married on 25 January 1946 at Tabernacle chapel in Aberystwyth. They then went to live with her parents at Edge Hill Road, joining her sister Lena and her husband Elfed who had married in 1942. On 25 November 1946 Owen and Jane's daughter Valerie was born.

Owen and Jane were then able to secure the tenancy of a new two-bedroomed flat built by the local council in Penparcau, and moved there to begin their new independent family life.

The railway recommenced in July 1945. Isaac Jenkins and his son Arthur were pleased to be active once more. There was a small hut beside the line, at about eight and three-quarter miles up the line. The wooden hut had a door and a small, shuttered hatch window. The guard on the train would pass down the pay tins to Isaac. He would enter the hut, open the hatch and formally pay his gang members through the hatch,

Isaac Jenkins and his son Arthur working on the track

Owen Jenkins working on his engine

116

even though he knew them all well and they worked alongside him.

Owen continued to work on the footplate of the engines, eventually becoming a driver. By 1959, Isaac's son Arthur succeeded him as Chief Ganger.

Despite the temporary closure of *Lein Fach* during the period 1939 to 1945, the railways nationally played an important role in the transport infrastructure of the country and the war effort. They transported troops and war materials to all parts of the country, but in the following years things changed, with government policies favouring the construction of new roads and motorways. So railway investment deteriorated. The rail system, especially that in Wales, was decimated by Dr Beeching in the early 1960s, resulting in a reduction of over 6,000 miles of track and isolating communities.

Isaac Jenkins retired from work on his beloved *Lein Fach* to continue to live at Gamlyn Cottage. He rejoiced in the fact that both his sons continued to support the railway; Arthur worked as a conscientious ganger on the line, and Owen drove the engine up and down the valley.

In testimony to the hard work and enthusiasm of Isaac Jenkins and his sons Arthur and Owen, as well as many others employed on the railway, the Vale of Rheidol railway line still continues today as a vibrant tourist attraction.

Epilogue

A new generation

The lives of a master mariner who became a farmer, and a lead miner who became a coal miner became interlinked through the relationships of their offspring. The need for gainful employment in rural communities, and the financial attraction of urban-based employment brought together people from different backgrounds. War further cemented the changes in both town and country.

The threads linking people across time and rural area brought together families whose past was deeply rooted in a land which had provided an array of employment opportunities. These work-related circumstances were frequently interdependent, permitting often isolated communities to grow and prosper. However, the march of modernity and improved transport systems, both on sea and on land, would bring about irreversible change. This was also hastened by the effects of two world wars.

*

Lead has linked my family's story, from early maritime transportation to the underground toil of my ancestors, to a narrow-gauge railway along the steep-sided Rheidol Valley. Aberystwyth and its hinterland were affected by lead, and the scars that remain in the countryside are a testament to a past which had a high dependency on the various ores. Ores which brought wealth to a few, bankruptcy to some and hardship to many.

The Cardiganshire hills are barriers against change. In them are old memories, old beliefs, habits and unaltered ways. They can vanish in mist and be blotted out by rain, yet sparkle in sunlight. Mines are now abandoned – what little remains of them that are left are now the habitats of owls, bats and ravens.

Only spoil heaps and rubble remain to indicate where previously there were men and machinery extracting, crushing and washing ores. The cacophony of noise is now replaced by silence. These remote upland regions radiate a mystical attraction that somehow compels us to return with no particular aim or purpose in view. The hills, blessed by the labour of centuries; it is sufficient merely to be there – in the land of lead.

The Old Mine

The released rock falls
Remorselessly reverberating
Ever downward
In the narrow stone shaft
Sped by gravity
To the final splash
Of accelerated arrival
At the waterlogged level.
The Old Man is long gone
His tympanic tapping
Replaced by a still silence
Along a sinister emptiness
In what was once the site
Of mineral seeking industry.
Now the land rests
Ravaged for eternity.

                              Brian Davies

Family timeline

1842, 15 March: Birth of Captain William James

1857, 19 August: Death of William James Senior in Nantes, aged 54

1863: Captain William James gains First Mate Certificate

1865: Captain William James gains his Master's Certificate

1870: Loss of SS *William and Mary*

1876: Loss of SS *Eigen* in Casablanca

1876: Birth of John David Davies

1880: SS *Clarissa* at Avilés and problem with cargo

1884, 19 September: Birth of Catherine Jenkins

1885: Birth of Marged Ann Davies (Morgan)

1886: Loss of SS *Clarissa*

1892, 19 October: Death of Mary Jane James (Evans), aged 57

1893: Marriage of Captain William James to Jane James

1895, 15 August: Birth of William Richard James

1898, 17 April: Birth of Lizzie Jenkins

1906, 21 December: Marriage of John Davies and Marged Ann

1917, 19 August: Death of Captain William James

1919, 21 September: Birth of William Elfed Davies

1921, 2 March: Birth of Helena James

1923, 4 April: Birth of Owen Jenkins

1942, March: Death of John David Davies

1946, 13 June: Birth of Brian Davies

1947, 26 May: Birth of Nigel Davies

1954, 26 October: Death of Jane James

1954, 10 November: Birth of Avril Davies

1957, 3 February: Death of Helena Jenkins

1960, June: Death of Catherine Ellen Jenkins

1960, 12 May: Death of Marged Ann Davies

1966, 23 February: Death of Isaac Jenkins
1967, 15 February: Death of Lizzie James
1968, 25 July: Death of William Richard James
1993, 21 September: Death of Owen Jenkins
2002, 15 January: Death of Elfed Davies
2002, 8 July: Death of Jane Jenkins and daughter Valerie in a
  road traffic accident
2004, 14 April: Death of Lena Davies
2009: Death of Nigel Davies

Bibliography

Davies, Elfed & Brian, *Salem Soldier* (2012).

Green, C, *The Vale of Rheidol Light Railway* (1986).

Lewis, W J, *Lead Mining in Wales* (1967).

Lewis, W J, *Born on a Perilous Rock* (1980).

The Advance of the Egyptian Expeditionary Force 1917–18 (1919).

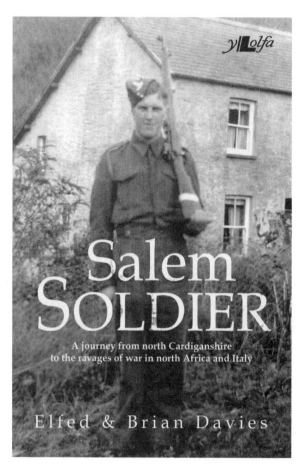

y Lolfa

Salem SOLDIER

A journey from north Cardiganshire
to the ravages of war in north Africa and Italy

Elfed & Brian Davies

£9.95

Also from Y Lolfa:

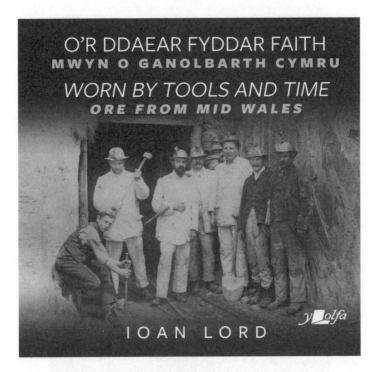

£14.99

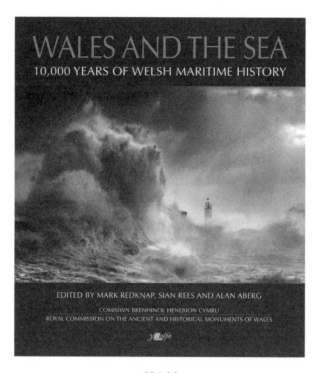

WALES AND THE SEA
10,000 YEARS OF WELSH MARITIME HISTORY

EDITED BY MARK REDKNAP, SIAN REES AND ALAN ABERG

COMISIWN BRENHINOL HENEBION CYMRU
ROYAL COMMISSION ON THE ANCIENT AND HISTORICAL MONUMENTS OF WALES

y Lolfa

£24.99